THE STORY-BOOK OF SCIENCE

ALSO AVAILABLE FROM LIVING BOOK PRESS

The Burgess Animal Book for Children (in color)

Home Geography
Elementary Geography
Viking Tales
Parables From Nature*
Fifty Famous Stories Retold*
The Blue Fairy Book*
     * in AO reading order

Richard Halliburton's Marvels of the Orient
Richard Halliburton's Marvels of the Occident

Charlotte Mason's Home Education Series

1. Home Education
2. Parents and Children
3. School Education
4. Ourselves
5. Formation of Character
6. A Philosophy of Education

And many, many more!

All Living Book Press titles are complete and unabridged, and presented with the original illustrations, sometimes from several sources, to bring these great books even more to life.

To see a complete list of all our releases or if you wish to leave us any feedback please visit www.livingbookpress.com

The Story-book of Science

Science

JEAN HENRI FABRE

LIVING BOOK
PRESS

This edition published 2018
by Living Book Press
Copyright © Living Book Press, 2018

Original edition published in 1917.

ISBN: 978-1-925729-25-2 (softback)
         978-1-922348-30-2 (hardback)

A catalogue record for this book is available from the National Library of Australia

CONTENTS

CHAPTER I

THE SIX

O NE EVENING, at twilight, they were assembled in a group, all six of them. Uncle Paul was reading in a large book. He always reads to rest himself from his labors, finding that after work nothing refreshes so much as communion with a book that teaches us the best that others have done, said, and thought. He has in his room, well arranged on pine shelves, books of all kinds. There are large and small ones, with and without pictures, bound and unbound, and even gilt-edged ones. When he shuts himself up in his room it takes something very serious to divert him from his reading. And so they say that Uncle Paul knows any number of stories. He investigates, he observes for himself. When he walks in his garden he is seen now and then to stop before the hive, around which the bees are humming, or under the elder bush, from which the little flowers fall softly, like flakes of snow; sometimes he stoops to the ground for a better view of a little crawling insect, or a blade of grass just pushing into view. What does he see? What does he observe? Who knows? They say, however, that there comes to his beaming face a holy joy, as if he had just found himself face to face with some secret of the wonders of God. It makes us feel better when we hear stories that he tells at these moments; we feel better, and furthermore we learn a number of things that some day may be very useful to us.

Uncle Paul is an excellent, God-fearing man, obliging to every one, and "as good as bread." The village has the greatest esteem for him, so much so that they call him Maître Paul, on account of his learning, which is at the service of all.

To help him in his field work—for I must tell you that Uncle

Paul knows how to handle a plow as well as a book, and cultivates his little estate with success—he has Jacques, the old husband of old Ambroisine. Mother Ambroisine has the care of the house, Jacques looks after the animals and fields. They are better than two servants; they are two friends in whom Uncle Paul has every confidence. They saw Paul born and have been in the house a long, long time. How often has Jacques made whistles from the bark of a willow to console little Paul when he was unhappy! How many times Ambroisine, to encourage him to go to school without crying, has put a hard-boiled new-laid egg in his lunch basket! So Paul has a great veneration for his father's two old servants. His house is their house. You should see, too, how Jacques and Mother Ambroisine love their master! For him, if it were necessary, they would let themselves be quartered.

Uncle Paul has no family, he is alone; yet he is never happier than when with children, children who chatter, who ask this, that, and the other, with the adorable ingenuousness of an awakening mind. He has prevailed upon his brother to let his children spend a part of the year with their uncle. There are three: Emile, Jules, and Claire.

Claire is the oldest. When the first cherries come she will be twelve years old. Little Claire is industrious, obedient, gentle, a little timid, but not in the least vain. She knits stockings, hems handkerchiefs, studies her lessons, without thinking of what dress she shall wear Sunday. When her uncle, or Mother Ambroisine, who is almost a mother to her, tells her to do a certain thing, she does it at once, even with pleasure, happy in being able to render some little service. It is a very good quality.

Jules is two years younger. He is a rather thin little body, lively, all fire and flame. When he is preoccupied about something, he does not sleep. He has an insatiable appetite for knowledge. Everything interests and takes possession of him. An ant drawing a straw, a sparrow chirping on the roof, are sufficient to engross his attention. He then turns to his uncle with his interminable questions: Why is this? Why is that? His uncle has great faith in this curiosity, which, properly guided, may lead to good results. But there is one thing about Jules that his uncle does not like. As we must be honest, we will own that Jules has a little fault which would become a grave one if not guarded against: he has a temper. If he is opposed he cries, gets angry, makes big eyes, and spitefully throws away his cap. But it is like the boiling over of milk soup: a trifle will calm him. Uncle Paul hopes to be able to bring him round by gentle reprimands, for Jules has a good heart.

Emile, the youngest of the three, is a complete madcap; his age permits it. If any one gets a face smeared with berries, a bump on the forehead, or a thorn in the finger, it is sure to be he. As much as Jules and Claire enjoy a new book, he enjoys a visit to his box of playthings. And what has he not in the way of playthings? Now it is a spinning-top that makes a loud hum, then blue and red lead soldiers, a Noah's Ark with all sorts of animals, a trumpet which his uncle has forbidden him to blow because it makes too much noise, then—But he is the only one that knows what there is in that famous box. Let us say at once, before we forget it, Emile is already asking questions of his uncle. His attention is awakening. He begins to understand that in this world a good top is not everything. If one of these days he should forget his box of playthings for a story, no one would be surprised.

CHAPTER II

FAIRY TALE AND THE TRUE STORY

THE SIX of them were gathered together. Uncle Paul was reading in a big book, Jacques braiding a wicker basket, Mother Ambroisine plying her distaff, Claire marking linen with red thread, Emile and Jules playing with the Noah's Ark. And when they had lined up the horse after the camel, the dog after the horse, then the sheep, donkey, ox, lion, elephant, bear, gazelle, and a great many others,—when they had them all arranged in a long procession leading to the ark, Emile and Jules, tired of playing, said to Mother Ambroisine: "Tell us a story, Mother Ambroisine—one that will amuse us."

And with the simplicity of old age Mother Ambroisine spoke as follows, at the same time twirling her spindle:

"Once upon a time a grasshopper went to the fair with an ant. The river was all frozen. Then the grasshopper gave a jump and landed on the other side of the ice, but the ant could not do this; and it said to the grasshopper: 'Take me on your shoulders; I weigh so little.' But the grasshopper said: 'Do as I do; give a spring, and jump.' The ant gave a spring, but slipped and broke its leg.

"Ice, ice, the strong should be kind; but you are wicked, to have broken the ant's leg—poor little leg.

"Then the ice said: 'The sun is stronger than I, and it melts me.'

"Sun, sun, the strong should be kind; but you are wicked, to melt the ice; and you, ice, to have broken the ant's leg—poor little leg.

"Then the sun said: 'The clouds are stronger than I; they hide me.'

"Clouds, clouds, the strong should be kind; but you are

4

wicked, to hide the sun; you, sun, to melt the ice; and you, ice, to have broken the ant's leg—poor little leg.

"Then the clouds said: 'The wind is stronger than we; it drives us away.'

"Wind, wind, the strong should be kind; but you are wicked, to drive away the clouds; you, clouds, to hide the sun; you, sun, to melt the ice; and you, ice, to have broken the ant's leg—poor little leg.

"Then the wind said: 'The walls are stronger than I; they stop me.'

"Walls, walls, the strong should be kind; but you are wicked, to stop the wind; you, wind, to drive away the clouds; you, clouds, to hide the sun; you, sun, to melt the ice; and you, ice, to have broken the ant's leg—poor little leg.

"Then the walls said: 'The rat is stronger than we; it bores holes through us.'

"Rat, rat, the strong—"

"But it is all the same thing, over and over again, Mother Ambroisine," exclaimed Jules impatiently.

"Not quite, my child. After the rat comes the cat that eats the rat, then the broom that strikes the cat, then the fire that burns the broom, then the water that puts out the fire, then the ox that quenches his thirst with the water, then the fly that stings the ox, then the swallow that snaps up the fly, then the snare that catches the swallow, then—"

"And does it go on very long like that?" asked Emile.

"As long as you please," replied Mother Ambroisine, "for however strong one may be, there are always others stronger still."

"Really, Mother Ambroisine," said Emile, "that story tires me."

"Then listen to this one: Once upon a time there lived a woodchopper and his wife, and they were very poor. They had seven children, the youngest so very, very small that a wooden shoe answered for its bed."

"I know that story," again interposed Emile. "The seven children are going to get lost in the woods. Little Hop-o'-my-Thumb marks the way at first with white pebbles, then with bread crumbs. Birds eat the crumbs. The children get lost, Hop-o'-my-Thumb, from the top of a tree, sees a light in the distance.

They run to it: rat-tat-tat! It is the dwelling of an ogre!"

"There is no truth in that," declared Jules, "nor in Puss-in-Boots, nor Cinderella, nor Bluebeard. They are fairy tales, not true stories. For my part, I want stories that are really and truly so."

At the words, true stories, Uncle Paul raised his head and closed his big book. A fine opportunity offered for turning the conversation to more useful and interesting subjects than Mother Ambroisine's old tales.

"I approve of your wanting true stories," said he. "You will find in them at the same time the marvelous, which pleases so much at your age, and also the useful, with which even at your age you must concern yourselves, in preparation for after life.

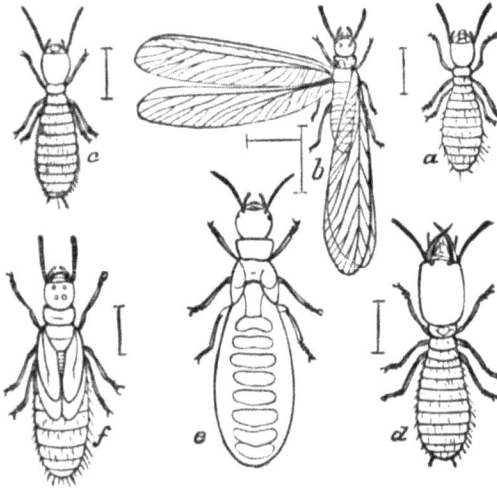

WHITE ANT

Believe me, a true story is much more interesting than a tale in which ogres smell fresh blood and fairies change pumpkins into carriages and lizards into lackeys. And could it be otherwise? Compared with truth, fiction is but a pitiful trifle; for the former is the work of God, the latter the dream of man. Mother Ambroisine could not interest you with the ant that broke its leg in trying to cross the ice. Shall I be more fortunate? Who wants to hear a true story of real ants?"

"I! I!" cried Emile, Jules, and Claire all together.

CHAPTER III

THE BUILDING OF THE CITY

"THEY ARE noble workers" began Uncle Paul, "Many a time, when the morning sun begins to warm up, I have taken pleasure in observing the activity that reigns around their little mounds of earth, each with its summit pierced by a hole for exit and entrance.

"There are some that come from the bottom of this hole. Others follow them, and still more, on and on. They carry between their teeth a tiny grain of earth, an enormous weight for them. Arrived at the top of the mound, they let their burden fall, and it rolls over the slope, and they immediately descend again into their well. They do not play on the way, or stop with their companions to rest a while. Oh! no: the work is urgent, and they have so much to do! Each one arrives, serious, with its grain of earth, deposits it, and descends in search of another. What are they so busy about?

"They are building a subterranean town, with streets, squares, dormitories, storehouses; they are hollowing out a dwelling-place for themselves and their family. At a depth where rain cannot penetrate they dig the earth and pierce it with galleries, which lengthen into long communicating streets, sub-divided into short ones, crossing one another here and there, sometimes ascending, sometimes descending, and opening into large halls. These immense works are executed grain by grain, drawn by strength of the jaws. If any one could see that black army of miners at work under the ground, he would be filled with astonishment.

"They are there by the thousands, scratching, biting, drawing, pulling, in the deepest darkness. What patience! What efforts! And when the grain of sand has at last given way, how they go off, head held high and proud, carrying it triumphantly above! I have seen ants, whose heads tottered under the tremendous load, exhaust themselves in getting to the top of the mound. In jostling their companions, they seemed to say: See how I work! And nobody could blame them, for the pride of work is a noble pride. Little by little, at the gate of the town, that is to say at the edge of the hole, this little mound of earth is piled up, formed by excavated material from the city that is being built. The larger the mound, the larger the subterranean dwelling, it is plain.

"Hollowing out these galleries in the ground is not all; they must also prevent landslides, fortify weak places, uphold the vaults with pillars, make partitions. These miners are then seconded by carpenters. The first carry the earth out of the ant-hill, the second bring the building materials. What are these materials! They are pieces of timber-work, beams, and small joists, suitable for the edifice. A tiny little bit of straw is a solid beam for a ceiling, the stem of a dry leaf can become a strong column. The carpenters explore the neighboring forests, that is to say the tufts of grass, to choose their pieces.

"Good! see this covering of an oat-grain. It is very thin, dry, and solid. It will make an excellent plank for the partition they are constructing below. But it is heavy, enormously heavy. The ant that has made the discovery draws backward and makes itself rigid on its six feet. No success: the heavy mass does not move. It tries again, all its little body trembling with energy. The oat-husk just moves a tiny bit. The ant recognizes its powerlessness. It goes off. Will it abandon the piece? Oh! no. When one is an ant, one has the perseverance that commands success. Here it is coming back with two helpers. One seizes the oat in front, the others hitch themselves to the side, and behold! it rolls, it advances; it will get there. There are difficult steps, but the ants they meet along the route will give them a shoulder.

"They have succeeded, not without trouble. The oat is at the entrance to the under-ground city. Now things become complicated; the piece gets awry; leaning against the edge of the hole, it cannot enter. Helpers hasten up. Ten, twenty unite their efforts without success. Two or three of them, engineers perhaps, detach themselves from the band, and seek the cause of this insurmountable resistance. The difficulty is soon solved: they must put the piece with the point at the bottom. The oat is drawn back a little, so that one end overhangs the hole. One ant seizes this end while the others lift the end that is on the ground, and the piece, turning a somersault, falls into the well, but is prudently held on to by the carpenters clinging to the sides. You may perhaps think, my children, that the miners mounting with their grain of earth would stop from curiosity before this mechanical prodigy? Not at all, they have not time. They pass with their loads of excavated material, without a glance at the carpenters' work. In their ardor they are even bold enough to slide under the moving beams, at the risk of being crippled. Let them look out! That is their affair.

"One must eat when one works so hard. Nothing creates an appetite like violent exercise. Milkmaid ants go through the ranks; they have just milked the cows and are now distributing the milk to the workers."

Here Emile burst out laughing. "But that is not really and truly so?" said he to his uncle. "Milkmaid ants, cows, milk! It is a fairy tale like Mother Ambroisine's."

Emile was not the only one to be surprised at the peculiar expressions Uncle Paul had used. Mother Ambroisine no longer turned her spindle, Jacques did not plait his wickers, Jules and Claire stared with wide-open eyes. All thought it a jest.

"No, my dears," said Uncle Paul. "I am not jesting; no. I have not exchanged the truth for a fairy tale. It is true there are milkmaid ants and cows. But as that demands some explanation, we will put off the continuation of the story until to-morrow."

Emile drew Jules off into a corner, and said to him in confidence: "Uncle's true stories are very amusing, much more so than Mother Ambroisine's tales. To hear the rest about those wonderful cows I would willingly leave my Noah's Ark."

CHAPTER IV

THE COWS

THE NEXT day Emile, when only half awake, began to think of the ants' cows. "We must beg uncle," said he to Jules, "to tell us the rest of his story this morning."

No sooner said than done: they went to look for their uncle.

"Aha!" cried he upon hearing their request, "the ants' cows are interesting you. I will do better than tell you about them, I will show them to you. First of all call Claire."

Claire came in haste. Their uncle took them under the elder bush in the garden, and this is what they saw:

The bush is white with flowers. Bees, flies, beetles, butterflies, fly from one flower to another with a drowsy murmur. On the trunk of the elder, amongst the ridges of the bark, numbers of ants are crawling, some ascending, some descending. Those ascending are the more eager. They sometimes stop the others on the way and appear to consult them as to what is going on above. Being informed, they begin climbing again with even more ardor, proof that the news is good. Those descending go in a leisurely manner, with short steps. Willingly they halt to rest or to give advice to those who consult them. One can easily guess the cause of the difference in eagerness of those ascending and those descending. The descending ants have their stomachs swollen, heavy, deformed, so full are they; those ascending have their stomachs thin, folded up, crying hunger. You cannot mistake them: the descending ants are coming back from a feast and, well fed, are returning home with the slowness that a heavy paunch demands; the ascending ants are running to the same feast and put into the assault of the bush the eagerness of an empty stomach.

"What do they find on the elder to fill their stomachs?" asked Jules. "Here are some that can hardly drag along. Oh, the gluttons!"

"Gluttons! no," Uncle Paul corrected him; "for they have a worthy motive for gorging themselves. There is above, on the elder, an immense number of the cows. The descending ants have just milked them, and it is in their paunch that they carry the milk for the common nourishment of the ant-hill colony. Let us look at the cows and the way of milking them. Don't expect, I warn you, herds like ours. One leaf serves them for pasturage."

Uncle Paul drew down to the children's level the top of a branch, and all looked at it attentively. Innumerable black velvety lice, immobile and so close together as to touch one another, cover the under side of the leaves and the still tender wood. With a sucker more delicate than a hair plunged into the bark, they fill themselves peacefully with the sap of the elder without changing their position. At the end of their back, they have two short and hollow hairs, two tubes from which, if you look attentively, you can see a little drop of sugary liquid escape from time to time. These black lice are called plant-lice. They are the ants' cows. The two tubes are the udders, and the liquor which drips from their extremity is the milk. In the midst of the herd, on the herd, even, when the cattle are too close together, the famished ants come and go from one louse to another, watching for the delicious little drop. The one who sees it runs, drinks, enjoys it, and seems to say on raising its little head: Oh, how good, oh, how good it is! Then it goes on its way looking for another mouthful of milk. But plant-lice are stingy with their milk; they are not always disposed to let it run through their tubes. Then the ant, like a milkmaid ready to milk her cow, lavishes the most endearing caresses on the plant-louse. With its antennæ, that is to say, with its little delicate flexible horns, it gently pats the stomach

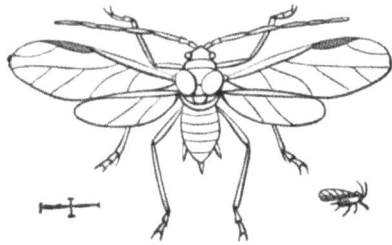

PLANT-LOUSE

and tickles the milk-tubes. The ant nearly always succeeds. What cannot gentleness accomplish! The plant-louse lets itself be conquered; a drop appears which is immediately licked up. Oh, how good, how good! As the little paunch is not full, the ant goes to other plant-lice trying the same caresses.

Uncle Paul let go the branch, which sprang back into its natural position. Milkmaids, cattle, and pasture were at once at the top of the elder bush.

"That is wonderful, Uncle," cried Claire.

"Wonderful, my dear child. The elder is not the only bush that nourishes milk herds for the ants. Plant-lice can be found on many other forms of vegetation. Those on the rosebush and cabbage are green; on the elder, bean, poppy, nettle, willow, poplar, black; on the oak and thistle, bronze color; on the oleander and nut, yellow. All have the two tubes from which oozes the sugary liquor; all vie with one another in feasting the ants."

Claire and her uncle went in-doors. Emile and Jules, enraptured by what they had just seen, began to look for lice on other plants. In less than an hour they had found four different kinds, all receiving visits of no disinterested sort from the ants.

CHAPTER V

THE SHEEPFOLD

IN THE evening Uncle Paul resumed the story of the ants. At that hour Jacques was in the habit of going the round of the stables to see if the oxen were eating their fodder and if the well-fed lambs were sleeping peacefully beside their mothers. Under the pretense of giving the finishing touches to his wicker basket, Jacques stayed where he was. The real reason was that the ants' cows were on his mind. Uncle Paul related in detail what they had seen in the morning on the elder: how the plant-lice let the sugary drops ooze from their tubes, how the ants drank this delicious liquid and knew how, if necessary, to obtain it by caresses.

"What you are telling us, Master," said Jacques, "puts warmth into my old veins. I see once more how God takes care of His creatures, He who gives the plant-louse to the ant as He gives the cow to man."

"Yes, my good Jacques," returned Uncle Paul, "these things are done to increase our faith in Providence, whose all-seeing eye nothing can escape. To a thoughtful person, the beetle that drinks from the depths of a flower, the tuft of moss that receives the rain-drop on the burning tile, bear witness to the divine goodness.

"To return to my story. If our cows wandered at will in the country, if we were obliged to take troublesome journeys to go and milk them in distant pastures, uncertain whether we should find them or not, it would be hard work for us, and very often impossible. How do we manage then? We keep them close at hand, in inclosures and in stables. This also is sometimes done by the ants with the plant-lice. To avoid tiresome journeys, sometimes useless, they put their herds in a park. Not all have this admirable foresight, however. Besides, if they had, it would be impossible to construct a park large enough for such innumerable cattle and their pasturage. How, for example, could they inclose in walls the willow that we saw this morning with its population of black lice? It is necessary to have conditions that are not beyond the forces available. Given a tuft of grass whose base is covered with a few plant-lice, the park is practicable.

"Ants that have found a little herd plan how to build a sheepfold, a summer châlet, where the plant-lice can be inclosed, sheltered from the too bright rays of the sun. They too will stay at the châlet for some time, so as to have the cows within reach and to milk them at leisure. To this end, they begin by removing a little of the earth at the base of the tuft so as to uncover the upper part of the root. This exposed part forms a sort of natural frame on which the building can rest. Now grains of damp earth are piled up one by one and shaped into a large vault, which rests on the frame of the roots and surrounds the stem above the point occupied by the plant-lice. Openings are made for the service of the sheepfold. The châlet is finished. Its inmates enjoy cool and quiet, with an assured supply of provisions. What more is needed for happiness? The cows are there, very peaceful, at their rack, that is to say, fixed by their suckers to the bark. Without leaving home the ants can drink to satiety that sweet milk from the tubes.

"Let us say, then, that the sheepfold made of clay is a building of not much importance, raised with little expense and hastily. One could overturn it by blowing hard. Why lavish such pains on so temporary a shelter? Does the shepherd in the high mountains take more care of his hut of pine branches, which must serve him for one or two months?

"It is said that ants are not satisfied with inclosing small herds of plant-lice found at the base of a tuft of grass, but that they also bring into the sheepfold plant-lice encountered at a distance. They thus make a herd for themselves when they do not find one already made. This mark of great foresight would not surprise me; but I dare not certify it, never having had the chance to prove it myself. What I have seen with my own eyes is the sheepfold of the plant-lice. If Jules looks carefully he will find some this summer, when the days are warmest, at the base of various potted plants."

"You may be sure, Uncle," said Jules, "I shall look for them. I want to see those strange ants' châlets. You have not yet told us why ants gorge themselves so, when they have the good luck to find a herd of plant-lice. You said those descending the elder with their big stomachs were going to distribute the food in the ant-hill."

"A foraging ant does not fail to regale itself on its own account if the occasion offers; and it is only fair. Before working for others must one not take care of one's own strength? But as soon as it has fed itself, it thinks of the other hungry ones. Among men, my child, it does not always happen so. There are people who, well fed themselves, think everybody else has dined. They are called egoists. God forbid your ever bearing that sorry name, of which the ant, paltry little creature, would be ashamed! As soon as it is satisfied, then, the ant remembers the hungry ones, and consequently fills the only vessel it has for carrying liquid food home; that is to say, its paunch.

"Now see it returning, with its swollen stomach. Oh! how it has stuffed so that others may eat! Miners, carpenters, and all the workers occupied in building the city await it so as to resume their work heartily, for pressing occupations do not permit them to go and seek the plant-lice themselves. It meets a carpenter, who for an instant drops his straw. The two ants meet mouth to mouth, as if to kiss. The milk-carrying ant disgorges a tiny little bit of the contents of its paunch, and the other one drinks the drop with avidity. Delicious! Oh! now how courageously it will work! The carpenter goes back to his straw again, the milk-carrier continues his delivery route. Another hungry one is met. Another kiss, another drop disgorged and passed from mouth to mouth. And so on with all the ants that present themselves, until the paunch is emptied. The milk-ant then departs to fill up its can again.

"Now, you can imagine that, to feed by the beakful a crowd of workers who cannot go themselves for victuals, one milk-ant is not enough; there must be a host of them. And then, under the ground, in the warm dormitories, there is another population of hungry ones. They are the young ants, the family, the hope of the city. I must tell you that ants, as well as other insects, hatch from an egg, like birds."

"One day," interposed Emile, "I lifted up a stone and saw a lot of little white grains that the ants hastened to carry away under the ground."

"Those white grains were eggs," said Uncle Paul, "which the ants had brought up from the bottom of their dwelling to expose them under the stone to the heat of the sun and facilitate their hatching. They hurried to descend again, when the stone was raised, so as to put them in a safe place, sheltered from danger.

"On coming out from the egg, the ant has not the form that you know. It is a little white worm, without feet, and quite powerless, not even able to move. There are in an ant-hill thousands of those little worms. Without stop or rest, the ants go from one to another, distributing a beakful, so that they begin to grow and change in one day into ants. I leave you to think how much they must work and how many plant-lice must be milked, merely to nurse the little ones that fill the dormitories."

CHAPTER VI

THE WILY DERVISH

"THERE ARE ant-hills everywhere, large or small," observed Jules. "Even in the garden I could have counted a dozen. From some the ants are so numerous they blacken the road when they come out. It must take a great many plant-lice to nourish all that little colony."

"Numerous though they be," his uncle assured him, "they will never lack cows, as plant-lice are still more numerous. There are so many that they often seriously menace our harvests. The miserable louse declares war against us. To understand it, listen to this story:

"There was once a king of India who was much bored. To entertain him, a dervish invented the game of chess. You do not know this game. Well, on a board something like a checkerboard two adversaries range, in battle array, one white, the other black, pieces of different values: pawns, knights, bishops, castles, queen and king. The action begins. The pawns, simple foot-soldiers, are destined as always to receive the first of the glory on the battlefield. The king looks on at their extermination, guarded by his grandeur far from the fray. Now the cavalry charge, slashing with their swords right and left; even the bishops fight with hot-headed enthusiasm, and the ambulating castles go here and

CHESS-BOARD WITH PIECES IN POSITION

there, protecting the flanks of the army. Victory is decided. Of the blacks, the queen is a prisoner; the king has lost his castles; one knight and one bishop do wonderful deeds to procure his flight. They succumb. The king is checkmated. The game is lost.

"This clever game, image of war, pleased the bored king very much, and he asked the dervish what reward he desired for his invention.

"'Light of the faithful,' answered the inventor, 'a poor dervish is easily satisfied. You shall give me one grain of wheat for the first square of the chessboard, two for the second, four for the third, eight for the fourth, and you will double thus the number of grains, to the last square, which is the sixty-fourth. I shall be satisfied with that. My blue pigeons will have enough grain for some days.'

"'This man is a fool,' said the king to himself; 'he might have had great riches and he asks me for a few handfuls of wheat.' Then, turning to his minister:—'Count out ten purses of a thousand sequins for this man, and have a sack of wheat given him. He will have a hundred times the amount of grain he asks of me.'

"'Commander of the faithful,' answered the dervish, 'keep the purses of sequins, useless to my blue pigeons, and give me the wheat as I wish.'

"'Very well. Instead of one sack, you shall have a hundred.'

"'It is not enough, Sun of Justice.'

"'You shall have a thousand.'

"'Not enough, Terror of the unfaithful. The squares of my chessboard would not have their proper amount.'

"In the meantime the courtiers whispered among themselves, astonished at the singular pretensions of the dervish, who, in the contents of a thousand sacks, would not find his grain of wheat doubled sixty-four times. Out of patience, the king convoked the learned men to hold a meeting and calculate the grains of wheat demanded. The dervish smiled maliciously in his beard, and modestly moved aside while awaiting the end of the calculation.

"And behold, under the pen of the calculators, the figure grew larger and larger. The work finished, the head one rose.

"'Sublime Commander,' said he, 'arithmetic has decided. To

satisfy the dervish's demand, there is not enough wheat in your granaries. There is not enough in the town, in the kingdom, or in the whole world. For the quantity of grain demanded, the whole earth, sea and continents together, would be covered with a continuous bed to the depth of a finger.'

"The king angrily bit his mustache and, unable to count out to him his grains of wheat, named the inventor of chess prime vizier. That is what the wily dervish wanted."

"Like the king, I should have fallen into the dervish's snare," said Jules. "I should have thought that doubling a grain sixty-four times would only give a few handfuls of wheat."

"Henceforth," returned Uncle Paul, "you will know that a number, even very small, when multiplied a number of times by the same figure, is like a snow-ball which grows in rolling, and soon becomes an enormous ball which all our efforts cannot move."

"Your dervish was very crafty," remarked Emile. "He modestly contented himself with one grain of wheat for his blue pigeons, on condition that they doubled the number on each square. Apparently, he asked next to nothing; in reality, he asked more than the king possessed. What is a dervish, Uncle?"

"In the religions of the East they call by that name those who renounce the world to give themselves up to prayer and contemplation."

"You say the king made him prime vizier. Is that a high office?"

"Prime vizier means prime minister. The dervish then became the greatest dignitary of the State, after the king."

"I am no longer surprised that he refused the ten purses of a thousand sequins. He was waiting for something better. The ten purses, however, would make a good sum?"

"A sequin is a gold piece worth about twelve francs. At that rate, the king offered the dervish a sum of one hundred and twenty thousand francs, besides the sacks of wheat."

"And the dervish preferred the grain sixty-four times doubled."

"In comparison what was offered him was nothing."

"And the plant-lice?" asked Jules.

"The story of the dervish is bringing us to that directly," his uncle assured him.

CHAPTER VII

A NUMEROUS FAMILY

"A PLANT-LOUSE, WE will suppose," resumed Uncle Paul, "has just established itself on the tender shoot of a rosebush. It is alone, all alone. A few days after, young plant-lice surround it: they are its sons. How many are there? Ten, twenty, a hundred? Let us say ten. Is that enough to assure the preservation of the species? Don't laugh at my question. I know well that if the plant-lice were missing from the rosebushes, the order of things would not be sensibly changed."

"The ants would be the most to be pitied," said Emile.

"The round earth would continue to turn just the same, even when the last plant-louse was dying on its leaf; but it is not, in truth, an idle question to ask if ten plant-lice suffice to preserve the race; for science has no higher object than the quest of providential means for maintaining everything in a just measure of prosperity.

"Well, ten plant-lice coming from one would be far too many if we did not have to take account of destructive agencies. One replacing one, the population remains the same; ten replacing one, in a short time the number increases beyond all possible limits. Think of the dervish's grain of wheat doubled sixty-four times, so that it becomes a bed of wheat of a finger's depth over the whole earth. What would it be if it had been multiplied ten times instead of doubled! In like manner, after a few years, the descendants of a first plant-louse, continually multiplied tenfold, would be in straitened circumstances in this world. But there is the great reaper, death, which puts an invincible obstacle to overcrowding, counterbalances life in its overgrowing fecundity, and, in partnership with it, keeps all things in a perpetual youth. On a rosebush apparently most peaceful there is death every minute. But the small, the humble, and weak, are the habitual pasture, the daily bread, of the large eaters. To how many dangers is not the plant-louse exposed, so tiny, so weak, and without any means of defense! No sooner does a little bird, hardly out of the shell, discover with its piercing eyes a spot haunted by the plant-lice, than, merely as an appetizer, it will swallow hundreds. And if a worm, far more rapacious, a horrible worm expressly created and put into the world to eat you alive, joins in, ah! my poor plant-lice, may God, the good God of little creatures, protect you; for your race is indeed in peril.

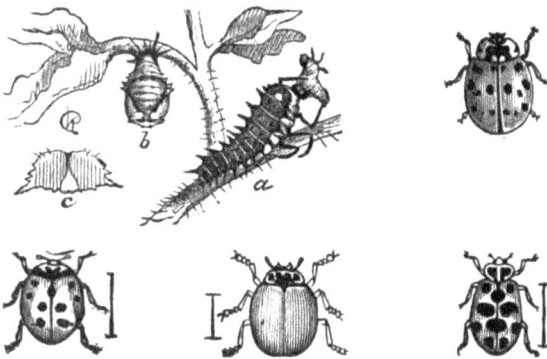

LADYBUG
(A) LARVA (B) PUPA (C) FIRST JOINT OF LARVA

"This devourer is of a delicate green with a white stripe on its back. It is tapering in front, swollen at the back. When it doubles itself up it takes the shape of a tear-drop. They call it the ants' lion because of the ravages it makes in the stupid herd. It establishes itself among them. With its pointed mouth, it seizes one, the biggest, the plumpest; it sucks it and throws away the skin, which is too hard for it. Its pointed head is lowered again, a second plant-louse seized, raised from the leaf, and sucked. Then another and another, a twentieth, a hundredth. The foolish herd, whose ranks are thinning, do not even seem to perceive what is going on. The trapped plant-louse kicks between the lion's fangs; the others, as if nothing were happening, continue to feed peacefully. It would take a good deal more than that to spoil their appetite! They eat while they are waiting to be eaten. The lion has had enough. He squats amidst the herd to digest at his ease. But digestion is soon over and already the greedy worm has its eye on those that he will soon crunch. After two weeks of continual feasting, after having browsed as it were on whole herds of plant-lice, the worm turns into an elegant little dragon-fly with eyes as bright as gold, and known as the hemerobius.

"Is that all? Oh, no. Here is the lady-bug, the good God's bug. It is round and red, with black spots. It is very pleasing; it has an innocent air. Who would take it also to be a devourer, filling its stomach with plant-lice? Look at it closely on the rosebush, and you will see it at its ferocious feasting. It is very pretty and innocent-looking; but it is a glutton, there is no denying the fact, so fond is it of plant-lice.

"Is that all? Oh, no. Those poor plant-lice are manna, the regular diet of all sorts of ravagers. Young birds eat them, the hemerobius eats them, lady-birds eat them, gluttons of all kinds eat them; and still there are always plant-lice. Ah! that is where, in the fight between fecundity which repairs and the rough battle of life which destroys, the weak excel by opposing legions and legions to the chances of annihilation. In vain the devourers come from all sides and pounce upon their prey; the devoured survive by sacrificing a million to preserve one. The weaker they are, the more fruitful they are.

"The herring, cod, and sardine are given over as pasturage for the devourers of the sea, earth, and sky. When they undertake long voyages to graze in favorable spots, their extermination is imminent. The hungry ones of the sea surround the school of fish; the famished ones of the sky hover over their route; those of the earth await them on the shore. Man hastens to lend a strong hand to the killing and to take his share of the sea food. He equips fleets, goes to the fish with naval armies in which all nations are represented; he dries in the sun, salts, smokes, packs. But there is no perceptible diminution in the supply; for him the weak are infinite in number. One cod lays nine million eggs! Where are the devourers that will see the end of such a family?"

"Nine million eggs!" exclaimed Emile. "Is that a great many?"

"Just to count them, one by one, would take nearly a year of ten working hours each day."

"Whoever counted them had lots of patience," was Emile's comment.

"They are not counted," replied Uncle Paul; "they are weighed, which is quickly done; and from the weight the number is deduced.

"Like the cod in the sea, the plant-lice are exposed on their rosebushes and alders to numerous chances of destruction. I have told you that they are the daily bread of a multitude of eaters. So, to increase their legions, they have rapid means that are not found in other insects. Instead of laying eggs, very slow in developing, they bring forth living plant-lice, which all, absolutely all, in two weeks have obtained their growth and begin to produce another generation. This is repeated all through the season, that is to say at least half the year, so that the number of generations succeeding one another during this period cannot be less than a dozen. Let us say that one plant-louse produces ten, which is certainly below the actual number. Each of these ten plant-lice borne by the first one bears ten more, making one hundred in all; each of these hundred bears ten, in all one thousand; each of the thousand bears ten, in all ten thousand; and so on, multiplying always by ten, eleven times. Here is the same calculation as the dervish's grain of wheat, which grew with such astonishing rapidity when they multiplied it by two. For the family of the plant-lice the increase is much more rapid, as the multiplication is made by ten. It is true that the calculation stops at the twelfth instead of going on to the sixty-fourth. No matter, the result would stupefy you; it is equal to a hundred thousand millions. To count a cod's eggs, one by one, would take nearly a year; to count the descendants of one plant-louse for six months would take ten thousand years! Where are the devourers that would see the end of the miserable louse? Guess how much space these plant-lice would cover, as closely packed as they are on the elder branch."

"Perhaps as large a place as our garden," suggested Claire.

"More than that; the garden is a hundred meters long and the same in width. Well, the family of that one plant-louse would cover a surface ten times larger; that is to say, ten hectares. What do you say to that? Is it not necessary that the young birds, little lady-bugs, and the dragon-fly with the golden eyes should work hard in the extermination of the louse, which if unhindered would in a few years overrun the world?

"In spite of the hungry ones which devour them, the plant-lice seriously alarm mankind. Winged plant-lice have been seen flying in clouds thick enough to obscure the daylight. Their black legions went from one canton to another, alighted on the fruit trees, and ravaged them. Ah! when God wishes to try us, the elements are not always unchained. He sends against us in our pride the paltriest of creatures. The invisible mower, the feeble plant-louse, comes, and man is filled with fear; for the good things of the earth are in great peril.

"Man, so powerful, can do nothing against these little creatures, invincible in their multitude."

Uncle Paul finished the story of the ants and their cows. Several times since, Emile, Jules, and Claire have talked of the prodigious families of the plant-louse and the cod, but rather lost themselves in the millions and thousand millions. Their uncle was right: his stories interested them much more than Mother Ambroisine's tales.

CHAPTER VIII

THE OLD PEAR-TREE

UNCLE PAUL had just cut down a pear-tree in the garden. The tree was old, its trunk ravaged by worms, and for several years it had not borne any fruit. It was to be replaced by another. The children found their Uncle Paul seated on the trunk of the pear-tree. He was looking attentively at something. "One, two, three, four, five," said he, tapping with his finger upon the cross-section of the felled tree. What was he counting?

"Come quick," he called, "come; the pear-tree is waiting to tell you its story. It seems to have some curious things to tell you."

The children burst out laughing.

"And what does the old pear-tree wish to tell us?" asked Jules.

"Look here, at the cut which I was careful to make very clean with the ax. Don't you see some rings in the wood, rings which begin around the marrow and keep getting larger and larger until they reach the bark?"

"I see them," Jules replied; "they are rings fitted one inside another."

"It looks a little like the circles that come just after throwing a stone into the water," remarked Claire.

"I see them too by looking closely," chimed in Emile.

"I must tell you," continued Uncle Paul, "that those circles are called annual layers. Why annual, if you please? Because one is formed every year; one only, understand, neither more nor less. The learned who spend their lives studying plants, and who are called botanists, tell us that no doubt is possible on that point. From the moment the little tree springs from the seed to the time when the old tree dies, every year there is formed a ring, a layer of wood. This understood, let us count the layers of our pear-tree."

Uncle Paul took a pin to guide his counting; Emile, Jules, and Claire looked on attentively. One, two, three, four, five—They counted thus up to forty-five, from the marrow to the bark.

"The trunk has forty-five layers of wood," announced Uncle Paul. "Who can tell me what that signifies? How old is the pear-tree?"

"That is not very hard," answered Jules, "after what you have just told us. As it makes one ring every year, and we have counted forty-five, the pear-tree must be forty-five years old."

"Eh! Eh! what did I tell you?" cried Uncle Paul, in triumph. "Has not the pear-tree talked? It has begun its history by telling us its age. Truly, the tree is forty-five years old."

"What a singular thing!" Jules exclaimed. "You can know the age of a tree as if you saw its birth. You count the layers of wood; so many layers, so many years. One must be with you, Uncle, to learn those things. And the other trees, oak, beech, chestnut, do they do the same?"

"Absolutely the same. In our country every tree counts one year for each layer. Count its layers and you have its age."

"Oh! how sorry I am I did not know that the other day," put in Emile, "when they cut down the big beech which was in the way on the edge of the road. Oh, my! What a fine tree! It covered a whole field with its branches. It must have been very old."

"Not very," said Uncle Paul. "I counted its layers; it had one hundred and seventy."

"One hundred and seventy, Uncle Paul! Honest and truly?"

"Honest and truly, my little friend, one hundred and seventy."

"Then the beech was a hundred and seventy years old," said Jules. "Is it possible? A tree to grow so old! And no doubt it would have lived many years longer if the road-mender had not had it cut down to widen the road."

"For us, a hundred and seventy years would certainly be a great age," assented his uncle; "no one lives so long. For a tree it is very little. Let us sit down in the shade. I have more to tell you about the age of trees."

CHAPTER IX

THE AGE OF TREES

"They used to tell of a chestnut of Sancerre whose trunk was more than four meters round. According to the most moderate estimate its age must have been three or four hundred years. Don't cry out at the age of this chestnut. My story is just beginning, and you may be sure that, as a narrator who stimulates the curiosity of his audience, I reserve the oldest for the end.

"Much larger chestnuts are known; for example, that of Neuve-Celle, on the borders of the Lake of Geneva, and that of Esaü, in the neighborhood of Montélimar. The first is thirteen meters round at the base of the trunk. From the year 1408 it sheltered a hermitage; the story has been testified to. Since then four centuries and a half have passed, adding to its age, and lightning has struck it at different times. No matter, it is still vigorous and full of leaves. The second is a majestic ruin. Its high branches are despoiled; its trunk, eleven meters round, is plowed with deep crevices, the wrinkles of old age. To tell the age of these two giants is hardly possible. Perhaps it might be reckoned at a thousand years, and still the two old trees bear fruit; they will not die."

"A thousand years! If Uncle had not said it, I should not believe it." This from Jules.

"Sh! You must listen to the end without saying anything," cautioned his uncle.

"The largest tree in the world is a chestnut on the slopes of Etna, in Sicily. Look at the map: you will see down there, at the extreme end of Italy, opposite the toe of that beautiful country which has the shape of a boot, a large island with three corners. That is Sicily. On that island is a celebrated mountain which throws up burning matter—a volcano, in short. It is called Etna. To come back to our chestnut, I must tell you that they call it 'the chestnut of a hundred horses,' because Jane, Queen of Aragon, visiting the volcano one day and, overtaken by a storm, took refuge under it with her escort of a hundred horsemen. Under its forest of leaves both riders and horses found shelter. To surround the giant, thirty people extending their arms and joining hands would not be enough. The trunk is more than fifty meters round. Judged by its size, it is less a tree-trunk than a fortress, a tower. An opening large enough to permit two carriages to pass abreast goes through the base of the chestnut and gives access into the cavity of the trunk, which is fitted up for the use of those who go to gather chestnuts; for the old colossus still has young sap and seldom fails to bear fruit. It is impossible to estimate the age of this giant by its size, for one suspects that a trunk as large as that comes from several chestnuts, originally distinct, but so near together that they have become welded into one.

"Neustadt, in Württemberg, has a linden whose branches, overburdened by years, are held up by a hundred pillars of masonry. The branches cover all together a space 130 meters in circumference. In 1229 this tree was already old, for writers of that time call it 'the big linden.' Its probable age to-day is seven or eight hundred years.

"There was in France, at the beginning of this century, an older tree than the veteran of Neustadt. In 1804 could be seen at the castle of Chaillé, in the Deux-Sèvres, a linden 15 meters round. It had six main branches propped

WHITE OAK

with numerous pillars. If it still exists it cannot be less than eleven centuries old.

"The cemetery of Allouville, in Normandy, is shaded by one of the oldest oaks in France. The dust of the dead, into which it has thrust its roots, seems to have given it an exceptional vigor. Its trunk measures ten meters in circumference at the base. A hermit's chamber surmounted by a little steeple rises in the midst of its enormous branches. The base of the trunk, partly hollow, is fitted up as a chapel dedicated to Our Lady of Peace. The greatest personages have esteemed it an honor to go and pray in this rustic sanctuary and meditate a moment under the shade of the old tree which has seen so many graves open and shut. According to its size, they consider this oak to be about nine hundred years old. The acorn that produced it must, then, have germinated about the year 1000. To-day the old oak carries its monstrous branches without effort. Glorified by men and ravaged by lightning, it peacefully follows the course of ages, perhaps having before it a future equal to its past.

"Much older oaks are known. In 1824 a wood-cutter of Ardennes felled a gigantic oak in whose trunk were found sacrificial vases and antique coins. The old oak had had fifteen or sixteen centuries of existence.

"After the Allouville oak I will tell you of some more companions of the dead; for it is above all in these fields of repose, where the sanctity of the place protects them against the injuries of man, that the trees attain such an advanced age. Two yews in the cemetery of Haie-de-Routot, department of Eure, merit attention above all. In 1832 they shaded with their foliage the whole of the field of the dead and a part of the church, without having experienced serious damage, when an extremely violent windstorm threw a part of their branches to the ground. In spite of this mutilation these two yews are still majestic old trees. Their trunks, entirely hollow, measure each of them nine meters in circumference. Their age is estimated at fourteen hundred years.

"That, however, is not more than half the age that some other trees of the same kind have attained. A yew in a Scotch cemetery measured twenty-nine meters around. Its probable age was two thousand five hundred years. Another yew, also in a cemetery in the same country, was, in 1660, so prodigious that the whole country was talking about it. They reckoned its age then at two thousand eight hundred and twenty-four years. If it is still standing, this patriarch of European trees bears the weight of more than thirty centuries.

"Enough for the present. Now it is your turn to talk."

"I like better to be silent, Uncle Paul," said Jules. "You have upset my mind with your trees that will not die."

"I am thinking of the old yew in the Scotch cemetery. Did you say three thousand years?" asked Claire.

"Three thousand years, my dear child; and we might go still further back, if I were to tell you of certain trees in foreign countries. Some are known to be almost as old as the world."

CHAPTER X

THE LENGTH OF ANIMAL LIFE

JULES AND Claire could not get over the astonishment caused by their uncle's story of the old trees to which centuries are less than years are to us. Emile, with his usual restlessness, led the conversation to another subject:

"And animals, Uncle," asked he, "how long do they live?"

"Domestic animals," was the reply, "seldom attain the age that nature allows them. We grudge them their nourishment, overtire them, and do not give them proper shelter. And then, we take from them their milk, fleece, hide, flesh, in fact everything. How can you ever grow old when the butcher is waiting for you at the stable door with his knife? Useless to speak of these poor victims of our need: to give us long life, they do not live out their time. Supposing that an animal is well treated, that it suffers neither hunger nor cold, that it lives in peace without excessive fatigue, without fear of knacker or butcher; under these good conditions, how many years will it live?

"Let us begin with the ox. Here is a robust one, I hope. What chest and shoulders! And then that big square forehead, with its vigorous horns around which the strap of the yoke goes; those eyes shining with the serene majesty of strength. If old age is the portion of the strong, the ox ought to live for centuries."

"I should think so too," assented Jules.

"Quite wrong, my dear children; the ox, so big, strong, massive, is old, very old, at twenty or thirty years. What to us would be verdant youth is for it decrepit old age.

"Let us pass on to the horse. You see I do not take my examples from among the weak; I choose the most vigorous. Well, the horse, as well as its modest companion, the ass, scarcely reaches more than thirty or thirty-five years."

"How mistaken I was!" Jules exclaimed. "I thought the horse and ox strong enough to live at least a century. They are so big, they take up so much room!"

"I do not know, my little friend, whether you can understand me, but I want to inform you that to take up a great deal of room in this world is not the way to live in peace and to enjoy a long life. There are people who take up a lot of space, not in the body—they are no bigger than we—but in their pretensions and their ambitious manœuvers. Do they live in peace, are they preparing for themselves a venerable old age? It is very doubtful. Let us remain small; that is to say, let us content ourselves with the little that God has given us; let us beware of the temptations of envy, the foolish counsels of pride; let us be full of activity, of work, and not of ambition. That is the only way we are permitted to hope for length of days.

"Let us return without delay to our animals. Our other domestic animals live a still shorter time. A dog, at twenty or twenty-five years, can no longer drag himself along; a pig is a tottering veteran at twenty; at fifteen at the most, a cat no longer chases mice, it says good-by to the joys of the roof and retires to some corner of a granary to die in peace; the goat and sheep, at ten or fifteen, touch extreme old age, the rabbit is at the end of its skein at eight or ten; and the miserable rat, if it lives four years, is looked upon among its own kind as a prodigy of longevity.

"Would you like me to tell you about birds? Very well. The pigeon may live from six to ten years; the guinea fowl, hen, and turkey, twelve. A goose lives longer; it is true that in its quality of goose it does not worry. The goose attains twenty-five years, and even a good deal more.

"But here is something better. The goldfinch, sparrow, birds free from care, always singing, always frisking, happy as possible with a ray of sunlight in the foliage and a grain of hemp-seed, live as long as the gluttonous goose, and longer than the stupid turkey. These very happy little birds live from twenty to twenty-five years, the age of an ox. As I told you, taking up a lot of room in this world is not the way to prepare oneself for a long life.

"As to man, if he leads a regular life, he often lives to eighty or ninety. Sometimes he reaches a hundred or even more. But should he attain only the ordinary age, the average age, as they say, that is about forty, then he is to be considered a privileged creature as to length of life; the foregoing facts show it. And besides, for man, my dear children, length of life is not measured exactly according to the number of years. He lives most who works most. When God calls us to Him, let us take with us the sincere esteem of others and the consciousness of having done our duty to the end; and, whatever our age, we shall have lived long enough."

CHAPTER XI

THE KETTLE

Now, THAT day, Mother Ambroisine was very tired. She had taken down from their shelves kettles, saucepans, lamps, candlesticks, casseroles, pans, and lids. After having rubbed them with fine sand and ashes, then washed them well, she had put the utensils in the sun to dry them thoroughly. They all shone like a mirror. The kettles particularly were superb with their rosy reflections; one might have said that tongues of fire were shining inside them. The candlesticks were a dazzling yellow. Emile and Jules were lost in admiration.

"I should like to know what they make kettles of, they shine so," remarked Emile. "They are very ugly outside, all black, daubed with soot; but inside, how beautiful they are!"

"You must ask Uncle," replied his brother.

"Yes," assented Emile.

No sooner said than done: they went in search of their uncle. He did not have to be entreated; he was happy whenever there was an opportunity to teach them something.

"Kettles are made of copper," he began.

"And copper?" asked Jules.

"Copper is not made. In certain countries, it is found already made, mixed with stone. It is one of the substances that it is not in the power of man to make. We use these substances as God has deposited them in the bosom of the earth for purposes of human industry; but all our knowledge and all our skill could not produce them.

"In the bosom of mountains where copper is found, they hollow out galleries which go down deep into the earth. There workmen called miners, with lamps to light them, attack the rock with great blows of the pick, while others carry the detached blocks outside. These blocks of stone in which copper is found are called ore. In furnaces made for the purpose they heat the ore to a very high temperature. The heat of our stove, when it is red-hot, is nothing in comparison. The copper melts, runs, and is separated from the rest. Then, with hammers of enormous weight, set in motion by a wheel turned by water, they strike the mass of copper which, little by little, becomes thin and is hollowed into a large basin.

"The coppersmith continues the work. He takes the shapeless basin and, with little strokes of the hammer, fashions it on the anvil to give it a regular shape."

"That is why coppersmiths tap all day with their hammers," commented Jules. "I had often wondered, when passing their shops, why they made so much noise, always tapping, without any stop. They were thinning the copper; shaping it into saucepans and kettles."

Here Emile asked: "When a kettle is old, has holes in it and can't be used, what do they do with it? I heard Mother Ambroisine speak of selling a worn-out kettle."

"It is melted, and another new kettle made out of the copper," replied Uncle Paul.

"Then the copper does not wear away?"

"It wears away too much, my friend: some of it is lost when they rub it with sand to make it shine; some is lost, too, by the continual action of the fire; but what is left is still good."

"Mother Ambroisine also spoke of recasting a lamp which had lost a foot. What are lamps made of?"

"They are of tin, another substance that we find ready-made in the bosom of the earth, without the power of producing it ourselves."

CHAPTER XII

METALS

"COPPER AND tin are called metals," continued Uncle Paul. "They are heavy, shining substances, which bear the blows of the hammer without breaking. They flatten, but do not break. There are still other substances which possess the considerable weight of copper and tin, as well as their brilliancy and resistance to blows. All these substances are called metals."

"Then lead, which is so heavy, is a metal too?" asked Emile.

"Iron also, silver and gold?" queried his brother.

"Yes, these substances and still others are metals. All have a peculiar brilliancy called metallic luster, but the color varies. Copper is red; gold, yellow; silver, iron, lead, tin, white, with a very slightly different shade one from another."

"The candlesticks Mother Ambroisine is drying in the sun," said Emile, "are a magnificent yellow and so shiny they dazzle. Are they gold?"

"No, my dear child; your uncle does not possess such riches. They are brass. To vary the colors and other properties of the metals, instead of always using them separately, they often mix two or three together, or even more. They melt them together, and the whole constitutes a sort of new metal, different from those which enter into its composition. Thus, in melting together copper and a kind of white metal called zinc, the same as the garden watering-pots are made of, they obtain brass, which has not the red of copper, nor the white of zinc, but the yellow of gold. The material of the candlesticks is, then, made of copper and zinc together; in a word, it is brass, and not gold, in spite of its luster and yellow color. Gold is yellow and glitters; but all that is yellow and glitters is not gold. At the last village fair they sold magnificent rings whose brilliancy deceived you. In gold, they would have cost a fine sum. The merchant sold them for a sou. They were brass."

"How can they tell gold from brass, since the color and luster are almost the same?" asked Jules.

"By the weight, chiefly. Gold is much heavier than brass; it is indeed the heaviest metal in frequent use. After it comes lead, then silver, copper, iron, tin, and finally zinc, the lightest of all."

"You told us that to melt copper," put in Emile, "they needed a fire so intense, that the heat of a red-hot stove would be nothing in comparison. All metals do not resist like that, for I remember very well in what a sorry way the first leaden soldiers you gave me came to their end. Last winter, I had lined them up on the luke-warm stove. Just when I was not watching, the troop tottered, sank down, and ran in little streams of melted lead. I had only time to save half a dozen grenadiers, and their feet were missing."

"And when Mother Ambroisine thoughtlessly put the lamp on the stove," added Jules, "oh! it was soon done for: a finger's breadth of tin had disappeared."

"Tin and lead melt very easily," explained Uncle Paul. "The heat of our hearth is enough to make them run. Zinc also melts without much trouble; but silver, then copper, then gold, and finally iron, need fires of an intensity unknown in our houses. Iron, above all, has excessive resistance, very valuable to us.

"Shovels, tongs, grates, stoves, are iron. These various objects, always in contact with the fire, do not melt, however; do not even soften. To soften iron, so as to shape it easily on the anvil by blows from the hammer, the smith needs all the heat of his forge. In vain would he blow and put on coal; he would never succeed in melting it. Iron, however, can be melted, but you must use the most intense heat that human skill can produce."

CHAPTER XIII

METAL PLATING

IN THE morning some wandering coppersmiths were passing. Mother Ambroisine had sold them the old kettle. Besides the sale, they were to make over the lamp whose foot had melted on the stove, and replate two saucepans. So the smiths lighted a fire in the open air, set up their bellows on the ground, and in a large round iron spoon melted the old lamp, adding a little tin to replace what had been lost. The melted metal was run into a mold, from which it came out in the shape of a lamp. This lamp, still pretty large, was fixed on a lathe which a little boy set in motion; and while it turned, the master touched it with the edge of a steel tool. The tin thus planed off fell in thin shavings, rolled up like curl-papers. The lamp was visibly becoming perfect; it took the proper polish and shape.

Afterward they busied themselves plating the copper saucepans. They cleaned them thoroughly inside with sand, put them on the fire, and, when they were very hot, went over the whole of their surface with a tow pad and a little melted tin. Wherever the pad rubbed, the tin stuck to the copper. In a few moments the inside of the saucepan, red before, was now shiny white.

Emile and Jules, while eating their little lunch of apples and bread, looked on at this curious work without saying a word. They promised themselves to ask their uncle the reason for whitening the inside of the copper saucepans with tin. In the evening, accordingly, they spoke of the tinning and plating.

"Highly cleaned and polished iron is very brilliant," explained their uncle. "The blade of a new knife, Claire's scissors, carefully kept in their case, are examples. But, if exposed to damp air, iron tarnishes quickly and covers itself with an earthy and red crust called—"

"Rust," interposed Claire.

"Yes, it is called rust."

"The big nails that hold the iron wires where the bell-flowers climb up the garden wall are covered with that red crust," remarked Jules; and Emile added:

"The old knife I found in the ground is covered with it too."

"Those large nails and the old knife are encrusted with rust because they have remained for a long time exposed to the air and dampness. Damp air corrodes iron; it becomes incorporated with the metal and makes it unrecognizable. When rusty, iron no longer has the properties that make it so useful to us; it is a kind of red or yellow earth, in which, without looking attentively, it would be impossible to suspect a metal."

"I can well believe it," said Jules. "For my part, I should never have taken rust for iron with which air and moisture had become incorporated."

"Many other metals rust like iron; that is to say, they are converted into earthy matter by contact with damp air. The color of rust varies according to the metal. Iron rust is yellow or red, that of copper is green, lead and zinc white."

"Then the green rust of old pennies is copper rust," said Jules.

"The white matter that covers the nozzle of the pump must be lead rust?" queried Claire.

"Exactly. The prime difficulty with rust is that it makes metals ugly: they lose their brilliance and polish; but it works still greater injury. There are harmless rusts which might get mixed with our food without danger: such is iron rust. On the contrary, copper and lead rusts are deadly poisons. If, by mischance, these rusts should get into our food, we might die, or at least we should experience cruel suffering. We will speak only of copper, for lead, on account of its quick melting, cannot go on the fire and is not used for kitchen utensils. Copper rust, I say, is a mortal poison; and yet they prepare food in copper vessels. Ask Mother Ambroisine."

"Very true," said she, "but I always have my eye on my saucepans: I keep them very clean and from time to time have them replated."

"I don't understand," put in Jules, "how the work that the tinsmith did this morning could prevent the copper rust being a poison."

"The smith's work will not make the copper rust cease to be a poison," replied Uncle Paul, "but it will prevent the rust's forming. Of the common metals tin rusts the least. Exposed to the air a long time, it scarcely tarnishes. And then the rust, which forms in small quantities, is innocuous, like iron rust. To prevent copper from covering itself with poisonous green spots, to preserve it from rust, it must be kept from contact with damp air and also with certain alimentary substances such as vinegar, oil, grease—substances that provoke the rapid formation of rust. For this reason the copper saucepan is coated over with tin inside. Under the thin bed of tin which covers it, the copper cannot rust, because it is no longer in contact with the air. The tin remains; but this metal changes with difficulty, and, besides, its rust, if it forms any, is harmless. So they plate copper, that is to say they cover it with a thin bed of tin, to prevent its rusting, and thus to prevent the formation of the dangerous poison that might, some day or other, be mixed with our food.

"They also tin iron, not to prevent the formation of poison, for the rust of this metal is harmless, but simply to preserve it from changing and covering itself with ugly red spots. This tinned iron is called tin-plate. Lids, coffee-pots, dripping-pans, graters, lanterns, and innumerable other things, are of tin-plate; that is to say, thin sheets of iron covered on both sides with a coating of tin."

CHAPTER XIV

GOLD AND IRON

"SOME METALS never rust; such a one is gold. Ancient gold pieces found in the earth after centuries are as bright as the day they were coined. No dross, no rust covers their effigy and inscription. Time, fire, humidity, air, cannot harm this admirable metal. Therefore gold, on account of its unchangeable luster and its rarity, is preeminently the material for ornaments and coins.

"Furthermore, gold is the first metal that man became acquainted with, long before iron, lead, tin, and the others. The reason why man's attention was called to gold, long centuries before iron, is not hard to understand. Gold never rusts; iron rusts with such grievous facility that in a short time, if we are not careful, it is converted into a red earth. I have just told you that gold objects, however old they may be, have come to us intact, even after having been in the dampest ground. As for objects of iron, not one has reached us that was not in an unrecognizable state. Corroded with rust, they have become a shapeless earthy crust. Now I will ask Jules if the iron ore that is extracted from the bowels of the earth can be real, pure iron, such as we use."

"It seems to me not, Uncle; for if iron at any given moment is pure, it must rust with time and change to earthy matter, as does the blade of a knife buried in the ground."

"My brother seems to reason correctly; I agree with him," said Claire.

"And gold?" Uncle Paul asked her.

"It is different with gold," she replied. "As that metal never rusts, is not changed by time, air, and dampness, it must pure."

"Exactly so. In the rocks where it is disseminated in small

scales, gold is as brilliant as in jewelers' boxes. Claire's earrings have not more luster than the particles set by nature in the rock. On the contrary, what a pitiful appearance iron makes when it is found! It is an earthy crust, a reddish stone, in which only after long research can one suspect the presence of a metal; it is, in fact, rust, mixed more or less with other substances. And then, it is not enough to perceive that this rusty stone contains a metal; a way must still be found to decompose the ore and bring the iron back to its metallic state. How many efforts were necessary to attain this result, one of the most difficult to achieve! How many fruitless attempts, how many painful trials! Iron, then, was the last to become of use to us, long after gold and other metals, like copper and silver, which are sometimes, but not always, found pure. That most useful of metals was the last; but with it an immense advance was made in human industry. From the moment man was in possession of iron, he found himself master of the earth.

"At the head of substances that resist shock, iron must be placed; and it is precisely its enormous resistance to rupture that makes this metal so precious to us. Never would a gold, copper, marble, or stone anvil resist the blows of the smith's hammer as an iron one does. The hammer itself, of what substance other than iron could it be made? If of copper, silver, or gold, it would flatten, crush, and become useless in a short time; for these metals lack hardness. If of stone, it would break at the first rather hard blow. For these implements nothing can take the place of iron. Nor can it for axes, saws, knives, the mason's chisel, the quarry-man's pick, the plowshare, and a number of other implements which cut, hew, pierce, plane, file, give or receive violent blows. Iron alone has the hardness that can cut most other substances, and the resistance that sets blows at defiance. In

HATCHET OF THE STONE AGE

this respect iron is, of all mineral substances, the handsomest present that Providence has given to man. It is preëminently the material for tools, indispensable in every art and industry."

"Claire and I read one day," said Jules, "that when the Spaniards discovered America, the savages of that new country had gold axes, which they very willingly exchanged for iron ones. I laughed at their innocence, which made them give such a costly price for a piece of very common metal. I think I see now that the exchange was to their advantage."

"Yes, decidedly to their advantage; for with an iron ax they could fell trees to make their dug-out canoes and their huts; they could better defend themselves against wild animals and attack the game in their hunts. This piece of iron gave them an assurance of food, a substantial boat, a warm dwelling, a redoubtable weapon. In comparison, a gold ax was only a useless plaything."

"If iron came last, what did men do before they knew of it?" asked Jules.

"They made their weapons and tools of copper; for, like gold, this metal is sometimes in a pure state so that it can be utilized just as nature gives it to us. But a copper implement, having little hardness, is of much less value than an iron one. Thus, in those far-off days of copper axes, man was indeed a wretched creature.

"He was still more so before knowing copper. He cut a flint into a point, or split it, and fastened it to the end of a stick; and that was his only weapon.

"With this stone he had to procure food, clothing, a hut, and to defend himself from wild beasts. His clothing was a skin thrown over his back, his dwelling a hut made of twisted branches and mud; his food a piece of flesh, produce of the chase. Domestic animals were unknown, the earth uncultivated, all industry lacking."

"And where was that?" asked Claire.

"Everywhere, my dear child; here, even in places where today are our most flourishing towns. Oh! how forlorn man was before attaining, by the help of iron, the well-being that we enjoy to-day; how forlorn was man and what a great present Providence made him in giving him this metal!"

Just as Uncle Paul finished, Jacques knocked discreetly at the door; Jules ran to open it. They whispered a few words to each other. It was about an important affair for the next day.

CHAPTER XV

THE FLEECE

A s was agreed upon the day before, Jacques made ready for the performance. To keep the patients from moving, they were obliged to make them lie down, their feet tied, between the two inclined planks of a rack. Steel knives shone on the ground. As for them, innocent victims of the needs of man, they were already bound and lying on their sides. With gentle resignation they awaited their sad fate. Were they going to be slain? Oh, no: they were to be shorn. Jacques took a sheep by its feet, placed it between the two planks of the rack, and, with large scissors, began, cra-cra-cra, to cut off the wool. Little by little, the fleece fell all in one piece. When the sheep had been despoiled, it ran free to one side, ashamed and chilly. It had just given its covering to clothe man. Jacques put another one on the rack, and the scissors began to move.

"Tell me, Jacques," said Jules, "are not the sheep very cold when they have had their wool cut off? See how that one trembles that you have just shorn."

"Never mind that: I have cho-sen a fine day for it. The sun is warm. By to-morrow they won't feel the need of their wool. And besides, ought not the sheep to suffer a little cold so that we may be warm?"

"We warm? How?"

"You astonish me. You do not know that, you who read so many books? Well, with this wool they will make you stockings and knitted things for this winter; they will even make cloth, fine cloth for clothes."

Spinning-wheel

"Peuh!" exclaimed Emile. "This wool is too dirty and ugly to make stockings, knitted things, and cloth."

"Dirty at present," Jacques agreed, "but it will be washed in the river, and when it has become very white Mother Ambroisine will work it on her spinning-wheel and make yarn of it. This yarn knitted with needles will become stockings that one is very glad to have on one's feet when obliged to run in the snow."

"I have never seen red, green, blue sheep; and yet there are red, green, blue, and other colored wools," said Emile.

"They dye the white wool that the sheep gives us; they put it into boiling water with drugs and coloring matter, and it comes out of that water with a color that stays."

"And cloth?"

"And cloth is made with threads of wool like those of stockings; but in order to weave these threads, make them cross each other regularly, and convert them into fabric, you must have complicated machines, weaving looms that cannot be had in our houses. These are only found in large factories used for manufacturing woolen goods."

"Then these trousers that I have on come from the sheep; this vest; my cravat, stockings too. I am dressed in the spoils of the sheep?" This from Jules.

"Yes, to defend ourselves from the cold, we take the sheep's wool. The poor beast furnishes its fleece for our clothes, its milk and flesh for our nourishment, its skin for our gloves. We live on the life of our domestic animals. The ox gives us his strength, flesh, hide; the cow, besides, gives us milk. The donkey, mule, horse, work for us. As soon as they are dead they leave us their skin, of which we make leather for our shoes. The hen gives us eggs, the dog puts his intelligence at our service. And yet there are people who, without any motive, maltreat these animals without which we should be so poor; who let them suffer hunger and beat them unmercifully! Never imitate those heartless ones; it would be an insult to God, who has given us the donkey, ox, sheep, and other animals. When I think that these valuable creatures give us all, even to their very life, I would share my last crust with them."

And the shears meanwhile continued their cra-cra-cra; and the fleece fell.

CHAPTER XVI

FLAX AND HEMP

W HILE LISTENING to what Jacques was saying about wool, Emile examined his handkerchief attentively. He turned it over and over, felt it, then looked through it. Jacques foresaw the question Emile was getting ready to ask him, and he said:

"Handkerchiefs and linens are not woolen. Certain plants, cotton, hemp, flax, and not sheep, furnish them; for, you see, I don't know much about those things myself. I have heard tell of the cotton plant, but have never seen it. And, besides, I am afraid talking to you will make me cut the sheep's skin."

In the evening, at Jules's request, they took up the history of the materials with which we clothe ourselves, and Uncle Paul explained their nature.

"The outside of hemp and flax is composed of long threads, very fine, supple, and tenacious, from which we manufacture our fabrics. We clothe ourselves with the spoils of the sheep, we make ourselves fine with the bark of the plant. The fabrics of luxury, cambric, tulle, gauze, point-lace, Mechlin lace, are made from flax; the stronger ones, even to coarse sacking, are of hemp. The cotton plant gives us the fabrics made of cotton.

"Flax is a slender plant with little delicate blue flowers, and is sown and harvested every year. It is much cultivated in Northern France, Belgium, and Holland. It is the first plant used by man for woven fabrics. Mummies of Egypt, the old land of Moses and the patriarchs, mummies which have lain buried

FLAX

46

four thousand years and more, are swathed in bands of linen."

"Mummies, did you say?" interposed Jules. "I don't know what they are."

"I will tell you, my dear child. Respect for the dead is found among all people and in all ages. Man regards as sacred what was the seat of a soul made in the image of God; he honors the dead, but the honors rendered differ according to time, place, customs. We inter the dead and put over the burial place a tombstone with an inscription, or at least a humble cross, divine emblem of life eternal. The ancients burned them on a funeral pile; they piously gathered the bones bleached by the fire and inclosed them in priceless vases. In Egypt, to preserve the cherished remains for the family, they embalmed the dead; that is to say, they impregnated them with aromatics and swathed them in linen to prevent decomposition. These pious duties were so delicately performed that, after centuries and centuries, we find intact in their chests of sweet-smelling wood, but dried and blackened by years, contemporaries of the ancient kings of Egypt, or the Pharaohs. These are what are called mummies.

"Hemp has been cultivated all over Europe for many centuries. It is an annual, of a strong, nauseous odor, with little, green, dull-looking flowers, whose stem, of the thickness of a quill pen, rises to about two meters. It is cultivated, like flax, both for its bark and for its grain, called hemp-seed."

"That is the grain, I think," said Emile, "we give the goldfinch, which it cracks with its beak when it breaks the shell to get out the little kernel."

"Yes, hemp-seed is the feast of little birds.

"The bark of the hemp has not the fineness of flax. The fibers of this latter plant are so fine that twenty-five grams of tow spun on the spinning-wheel furnishes a thread almost a league long. The spider's web alone can rival in delicacy certain linen fabrics.

"When hemp and flax reach maturity, they are harvested, and the seeds are separated by thrashing. The next operation, retting, then takes place, its purpose being to render the filaments of the bark, or the fibers, as they are called, easily separable from the wood. These fibers, in fact, are pasted to the stem and stuck together by a gummy substance that is very resistant and prevents separation until it is destroyed by rot. They sometimes do this retting by spreading the plants in the fields for a couple of weeks and turning them over now and then, until the tow detaches itself from the woody part or hemp-stalk.

"But the quickest way is to tie the flax and hemp in bundles and keep them submerged in a pond. There soon follows a rot which gives out intolerable smells; the bark decays, and the fiber, endowed with exceptional resistance, is freed.

"Then the bundles are dried; after that they crush them between the jaws of an instrument called a brake, to crush the stems into small pieces and separate the tow. Finally, to purge the tow of all woody refuse and to divide it into the finest threads, they pass it between the iron teeth of a sort of big comb called a heckle. In this state, the fiber is spun either by hand or by machine. The thread obtained is ready for weaving.

"On a loom they place in order, side by side, numerous threads composing what they call the warp. By turns, impelled by a pedal on which the operator's foot presses, one half of these threads descends while the other half ascends. At the same time the operator passes a transverse thread in a shuttle through the two halves of the warp, from left to right, then from right to left. From this inter-crossing comes the woven fabric. And it is finished; the garb of the plant has changed masters; the bark of the hemp has become cloth, that of flax a princely lace worth some hundreds of francs by the piece."

CHAPTER XVII

COTTON

"COTTON, THE most important of the materials used for our woven fabrics, is furnished by a semi-tropical plant called the cotton plant. It is an herb or even a shrub from one to two meters high, and its large yellow flowers are followed by an abundant fruitage of bolls, each as large as an egg, filled with a silky flock, sometimes brilliantly white, sometimes a pale yellowish shade, according to the kind of cotton. In the middle of this flock are the seeds."

"It seems to me I have seen flock of that kind fall in flakes in the spring from the top of poplars and willows," said Claire.

"The comparison is very good. Willows and poplars have for their fruit tiny little long and pointed bolls three or four times as large as a pin's head. In the month of May these bolls are ripe. They open and set free a very fine white down, in the middle of which are the seeds. If the air is calm, this down piles up at the foot of the tree in a bed of cotton wool, as white as snow; but at the least breath of wind the flakes are borne long distances, carrying with them the seeds, which thus find unoccupied places where they can germinate and become trees. Many other seeds are provided with soft aigrettes, silky plumes, which keep them up in the air a long time and permit them distant journeys in order to disseminate the plant. For example, who is not familiar with the seeds of thistles and dandelions, those beautiful silky plumed seeds that you take pleasure in blowing into the air?"

"Can the flock of poplar bolls be put to the same use as cotton?" Jules asked.

"By no means. There is too little of it, and it would be too difficult to gather. Besides, it is so short it might not be possible to spin it. But if we ourselves cannot make use of it, others find it very useful. This flock is the little birds' cotton; many gather it to line their nests. The goldfinch, among others, is one of the cleverest of the clever. Its house of cotton is a masterpiece of elegance and solidity. In the fork of several little branches, with the cottony flock of the willow and poplar, with bits of wool that hedge thorns pull out from sheep as they pass, with the plumy aigrettes of thistle seeds, it makes for its young a cup-shaped mattress, so soft and warm and wadded that no little prince in his swaddling-clothes ever had the like.

"To build their nests, birds find materials near at hand; they only have to set to work. When spring comes, the goldfinch does not have to think of the materials for its nest; it is sure that the osier-beds, thistles, and roadside hedges will furnish in abundance all that it needs. And it ought to be thus, for a bird has not the intelligence to prepare a long time in advance, by careful and wise industry, the things that it will need. Man, whose noble prerogative it is to acquire everything by work and reflection, procures cotton from distant countries; a bird finds its cotton on the poplars of its grove.

PICKING COTTON BY HAND

"At maturity the cotton bolls open wide, and their flock bursts out in soft flakes that are gathered by hand, boll by boll. The flock, well dried in the sun on screens, is beaten with flails or, better, submitted to the action of certain machines. It is thus freed from all seeds and husks. Without any other preparation, cotton comes to us in large bales to be converted into fabrics in our manufactories. The countries that furnish the most of it are India, Egypt, Brazil, and, above all, the United States of North America.

"In a single year the European manufactories work up nearly eight hundred million kilograms of cotton. This enormous weight is not too much, for the whole world clothes itself with the precious flock, turned into print, percale, calico. Thus human activity has no greater field than the cotton trade. How many workmen, how many delicate operations, what long voyages, all for a simple piece of print costing a few centimes! A handful of cotton is gathered, we will suppose, two or three thousand leagues from here. This cotton crosses the ocean, goes a quarter round the globe, and comes to France or England to be manufactured. Then it is spun, woven, ornamented with colored designs, and, converted into print, crosses the seas again, to go perhaps to the other end of the world to serve as head-dress for some woolly-haired negro. What a multiplicity of interests are brought into play! It was necessary to sow the plant; then, for a good half of the year, to cultivate it. Out of a handful of flock, then, provision must be made for the remuneration of those who have cultivated and harvested. Next come the dealer who buys and the mariner who transports it. To each of them is due a part of the handful of flock. Then follow the spinner, weaver, dyer, all of whom the cotton must indemnify for their work. It is far from being finished. Now come other dealers who buy the fabrics, other mariners who carry them to all parts of the world, and finally merchants who sell them at retail. How can the handful of flock pay all these interested ones without itself acquiring an exorbitant price?

"To accomplish this wonder two industrial powers intervene: work on a large scale and the aid of machinery. You have seen how Ambroisine spins wool on the wheel. The carded wool is first divided into long locks. One of these locks is applied to a hook which turns rapidly. The hook seizes the wool and in its rotation twists the fibers into one thread, which lengthens little by little at the expense of the lock held and regulated by the fingers. When the thread attains a certain length, Mother Ambroisine rolls it on the spindle by a suitable movement of the wheel; then she continues twisting the wool again.

"Strictly speaking, cotton could be spun in the same way; but, however clever Mother Ambroisine may be, the fabrics made from the thread of her wheel would cost an enormous price on account of the time spent. What, then, is to be done? A machine is made to spin the cotton. In rooms larger than the biggest church are placed, by hundreds of thousands, the nicely adjusted machines proper for spinning, with hooks, spindles, and bobbins. And all turn at the same time with a precision and rapidity that defy watching. The work goes on with noise enough to deafen you. The flock of cotton is seized by thousands and thousands of hooks; the endless threads come and go from one bobbin to another, and roll themselves on the spindles. In a few hours a mountain of cotton is converted into thread, the length of which would go several times around the whole earth. What have they spent for work which would have exhausted the strength of an army of spinners as clever as Mother Ambroisine? Some shovelfuls of coal to heat the water, the steam of which starts the machine that sets everything going. Weaving, the printing of the colored designs,—in short, the various operations that the flock undergoes to become cloth are executed by means quite as expeditious, quite as economical. And it is thus that the planter, broker, mariner, spinner, weaver, dyer, and merchant can all have their share in the handful of cotton flock which has become a piece of calico and is sold for four sous."

CHAPTER XVIII

PAPER

MOTHER AMBROISINE called Claire. A friend had just come to see her to learn about an embroidery stitch that troubled her. At the request of Jules and Emile, however, Uncle Paul continued. He knew Jules would take pleasure in repeating the conversation to his sister.

"Flax, hemp, and cotton, especially the last-named, have still another use of great importance. First they clothe us; then, when too ragged to use any more, they serve to make paper."

"Paper!" exclaimed Emile.

"Paper, real paper, that on which we write, of which we make books. The beautiful white sheets of your copybooks, the leaves of a book, even the costliest, gilt-edged and enriched with magnificent pictures, come to us from miserable rags.

"Despicable tatters are collected: some of them are picked up from the filth of the street, some are unspeakably filthy. They are sorted over, these for fine paper, those for coarse. They are thoroughly washed, for they need it. Now machines take them in hand. Scissors cut them, steel claws tear them, wheels make pulp of them and reduce them to shreds. Mill-stones take them and grind them still more, then triturate them in water, and convert them into a sort of soup. The pulp is gray, it must be whitened. Then recourse is had to powerful drugs, which attack everything they touch, and in less than no time make it white as snow. Behold the pulpy mass thoroughly purified. Other machines spread it in thin layers on sieves. Water drips through, and the rag soup forms into felt. Cylinders press this felt, others dry it, others give it a polish. The paper is finished.

"Before it became paper, the first material was rags, or cloth too tattered to use. How many uses has not this cloth served, and what energetic treatments has it not undergone before being cast out as rubbish! Washing with corrosive ashes, contact with acrid soap, pounding with a beetle, exposure to the sun, air, and rain. What is then this material which, in spite of its delicacy, resists the brutalities of washing, soap, sun, and air; which remains intact in the bosom of rottenness; which braves the machines and drugs of paper-making, and always comes out of these ordeals more supple and whiter, to become at last a sheet of paper, beautiful satiny paper, the confidant of our thoughts? You know now, my little friends, this admirable material, source of so much intellectual progress, comes to us from the flock of the cotton plant and the bark of hemp and flax."

"I am certainly going to surprise Claire," said Jules, "when I tell her that her beautiful prayer-book with the silver clasp was made from horrid rags, perhaps from ragged handkerchiefs thrown away for rubbish, or from tatters picked up from the mud of the street."

"Claire will be interested to learn the nature of paper; but, I am sure, the lowly origin of her prayer-book will not lessen the value of it in her mind. Skill performs a marvel in transforming despicable rags into a book, depository of noble thoughts. God, my dear child, does incomparably more in the miracle of vegetation. The filth of the dung-hill, when buried in the soil, becomes transformed into the most pleasing things in the world; for it becomes the rose, the lily, and other flowers. As for us, let us be like Claire's book and the flowers of the good God: let us try to have real value in ourselves, and let us never blush at our humble extraction. There is only one true greatness, only one true nobility: greatness and nobility of the soul. If we possess them, the merit is all the greater by reason of our lowly origin."

CHAPTER XIX

THE BOOK

"Now that I know what paper is made of," said Jules, "I should like to know how they make books."

"I could listen all day without getting tired," Emile asserted. "For a story I would leave my top and my soldiers."

"To make a book, my children, there is double work: first the labor of the one who thinks and writes it, then the labor of the one who prints it. To think a book and write it under the sole dictation of one's mind is a difficult and serious business. Brain-work exhausts our strength much more quickly than manual labor, for we must put the best of ourselves into it, our soul. I tell you these things that you may see what gratitude you owe those who, solicitous for your future, think and write in order to teach you to think for yourselves and to free you from the miseries of ignorance."

"I am quite convinced," returned Jules, "of the difficulties to be overcome in order to compose a book under the sole dictation of one's mind; for when I want to write a letter of half a page to wish you a Happy New Year, I come to a full stop at the first word. How hard it is to find the first word! My head is heavy, my face flushes, and I can't see straight. I shall do better when I know my grammar well."

"I am sorry, my dear child, but I must undeceive you. Grammar cannot teach one to write. It teaches us to make a verb agree with its subject, an adjective with a substantive, and other things of that kind. It is very useful, I admit, for nothing is more displeasing than to violate the rules of language; but that does not impart the gift of writing. There are people whose memories are crammed with rules of grammar, who, like you, stop short at the first word.

"Language is in some sort the clothing of thought. We cannot clothe what does not exist; we cannot speak or write what we do not find in our minds. Thought dictates and the pen writes. When the head is furnished with ideas, and usage, still more than grammar, has taught us the rules of language, we have all that is necessary to write excellent things correctly. But, again, if ideas are wanting, if there is nothing in the head, what can you write? How are these ideas to be acquired? By study, reading, and conversation with people better instructed than we."

"Then, in listening to all these fine things you tell us, I am no doubt learning to write," said Jules.

"Why, certainly, my little friend. Is it not true, for example, that if it had been proposed to you, a few days ago, to write only two lines about the origin of paper, you would not have been able to do it? What was wanting? Ideas and not grammar, although you know very little of that yet."

"It is true, I was entirely ignorant what paper comes from. To-day I know that cotton is a flock found in the bolls of a shrub called the cotton plant: I know that with this flock they make thread; then, after the thread, cloth; I know that when the cloth gets old with use, it is reduced to pulp by machines, and that this pulp, stretched in very thin layers and pressed, finally becomes a sheet of paper. I know these things well, and yet I should find it very hard to write them."

"You are mistaken, for all you need do is to put in writing exactly what you have just told me."

"You write then just as you talk?" asked the boy, incredulously.

"Yes, provided that speech is corrected, if necessary, on reflection, since writing gives time for it, whereas talking does not."

"In that case, I should soon have my five lines on paper. I should write: 'Cotton is a flock that is found in the bolls of a shrub called the cotton plant. With this flock they make thread; and with this thread, cloth. When the cloth is worn out, machines tear it into little pieces, and mill-stones grind it with water to make it into a pulp. This pulp is stretched in thin layers which are pressed and dried. Then it is paper.' There! Is that right, Uncle?"

"As well as one could wish from one of your age," his uncle

assured him.

"But that could not be put into a book."

"And why not? I promise you that shall be in a book some day. It has been said to me that our talks might be useful to many other little boys as desirous to learn as you, and I propose to collect them in all their simplicity and make a book of them."

"A book where I could read at leisure the stories that you tell us? Oh, how pleased I am, Uncle, and how I love you! You won't put my ignorant questions in that book?"

"I shall put them all in. You know next to nothing now, my dear child, but you ardently desire to learn. That is a fine quality, and a very becoming one."

"Are you at least sure that the little boys who read this book will not laugh at me?"

"I am sure."

"Tell them then that I love them well and embrace them all."

"Tell them I wish them as good a top and as fine lead soldiers as those you gave me," put in Emile.

"Take care, Emile," cautioned his brother. "Uncle may put your lead soldiers in the book."

"They will be there, they are there."

CHAPTER XX

PRINTING

"AFTER A book is written, the author sends his work, his manuscript, to the printer, who is to reproduce it in printed letters and in as many copies as are desired.

"Picture to yourself fine and short metal sticks, on the end of each of which is carved in relief a letter of the alphabet. One of these sticks has an a on the end, another a b, another a c, etc. There are others which have a full-stop, a comma, a semi-colon; in fact, there are as many distinct kinds of these little metal pieces as there are letters and orthographic signs in our written language. Besides, each letter and each sign are represented a great many times. Let us take note, too, that all these characters are carved wrong side before; you will soon see the reason.

"A workman called a compositor has before him a stand of cases, of which each compartment is occupied by a single letter of the alphabet, or by an orthographic sign. The a's are in such a compartment, the b's in a second, the c's in a third, and so on. The letters, furthermore, are not arranged in the case alphabetically. To shorten the work, they put in the compartments near to hand the letters that occur most frequently, such as the e's, r's, i's, a's; and they place in the more distant compartments the letters less often used, such as x's and y's.

"The compositor has before him a manuscript, and at his left hand a little flanged iron ruler called a composing-stick. As he reads, his right hand, guided by long habit, searches in the case the desired letter and places it in the composing-stick, upright and in a row with the others. He separates the words by means of a metal stick like those of the letters, but the end of which remains depressed and does not bear any carving. The first line

finished, the composi-
tor begins another by
setting a new row of
little metal pieces next
to the row already fin-
ished. Finally, when
the composing-stick is
full, the workman cau-
tiously places the con-
tents in an iron frame,
which keeps the deli-
cate combination from
going to pieces; and he

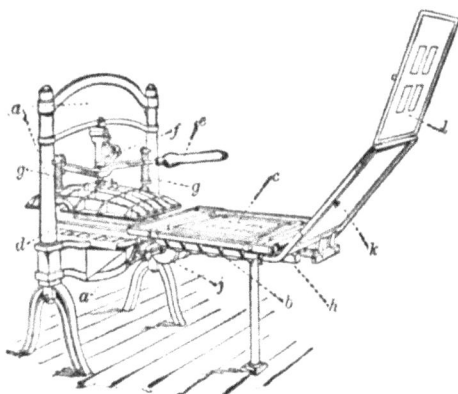

AN OLD FASHIONED HAND PRESS

continues thus until the frame is quite full and we have what is
called the printing-bed. This plate is composed of a multitude
of little metal sticks, simply placed side by side. There are as
many of these as there are letters, orthographic signs, and spaces
separating the words. The arrangement of these numerous bits
of metal is a masterpiece that a false movement might ruin. It
is held firm in its iron frame by means of wedges, so that the
whole thing seems made of a single block of metal. The bed is
then ready for printing.

"A roller impregnated with a thick ink made of oil and lamp-
black is passed over the plate. The letters and orthographic signs,
which alone stand out in relief, become covered with ink; the
rest does not take it because its surface is lower. A sheet of paper
is placed on the inked plate; it is covered with a pad to protect
it, then pressed hard. The ink of the characters is deposited on
the paper, and the sheet is found printed on one side. To print
the other, the operation is repeated with a second plate. The
metal letters are, as I said, carved wrong side before, as the letters of a book appear when you look at them in a mirror. The
inky imprint left by them on the paper reproduces them in a
reversed position, and consequently in the right way.

"The first sheet is followed immediately by a second. With
the roller the plate is inked again, a sheet of paper is applied,
pressure is exerted, and it is done. Then comes a third sheet, a
hundredth, a thousandth, indefinitely. All that is needed each

time is to ink the plate, cover it with paper, then press. All this is done with such rapidity that in a short time we have a great pile of printed sheets, each of which it would take a whole day to write by hand.

"Before the invention of this marvelous art, which enables us to reproduce the works of the mind very rapidly and in as great numbers as may be desired, we were restricted to hand-made copies. These manuscript books required years of work, and hence were very rare and high-priced. Large fortunes were necessary to acquire a library of several volumes. To-day books find their way everywhere, spreading in profusion, even among the lowest classes, the sacred bread of intelligence. Printing has been known for four hundred years: its invention is due to Gutenberg."

"That is a name I shall never forget," said Jules.

"It deserves, above all, to be remembered, for with the printed book Gutenberg rendered impossible henceforth the ignorant times through which man has miserably passed. Our intellectual treasures, resources for the future, are better than engraved on stone or metal; they are inscribed on sheets of paper, in copies too numerous to be all destroyed."

CHAPTER XXI

BUTTERFLIES

O H, HOW beautiful! Oh, my goodness, how beautiful they are!
There are some whose wings are barred with red on a gar-
net background; some bright blue with black circles; others are
sulphur-yellow with orange spots; again others are white fringed
with gold-color. They have on the forehead two fine horns, two
antennæ, sometimes fringed like an aigrette, sometimes cut off
like a tuft of feathers. Under the head they have a proboscis,
a sucker as fine as a hair and twisted into a spiral. When they
approach a flower, they untwist the proboscis and plunge it to
the bottom of the corolla to drink a drop of honeyed liquor. Oh,
how beautiful they are! Oh, my goodness, how beautiful they
are! But if one manages to touch them, their wings tarnish and
leave between the fingers a fine dust like that of precious metals.

Now their uncle told the children the names of the butterflies that flew on the flowers in the garden. "This one," said he,
"whose wings are white with a black border and three black
spots, is called the cabbage butterfly. This larger one, whose
yellow wings barred with black terminate in a long tail, at the
base of which are found a large rust colored eye and blue spots,
is called the swallow-tail. This tiny one, sky-blue above, silver-
gray underneath, sprinkled with black eyes in white circles, with
a line of reddish spots bordering the wings, is called the Argus."

And Uncle Paul continued thus, naming the butterflies that
a bright sun had drawn to the flowers.

"The Argus ought to be difficult to catch," observed Emile.
"He sees everywhere; his wings are covered with eyes."

"The pretty round spots that a great many butterflies have
on their wings are not really eyes, although they are called by

that name; they are ornaments, nothing more. Real eyes, eyes for seeing, are in the head. The Argus has two, neither more nor fewer than the other butterflies."

"Claire tells me," said Jules, "that butterflies come from caterpillars. Is it true, Uncle?"

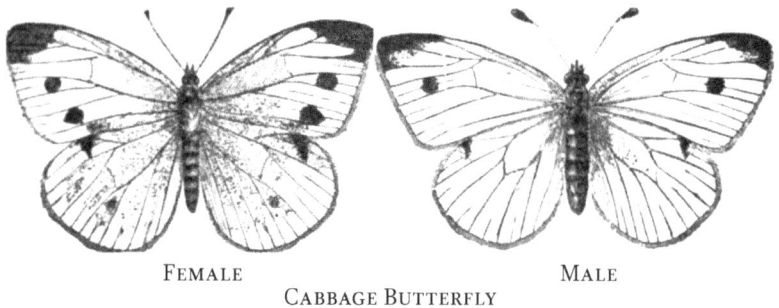

FEMALE   MALE
CABBAGE BUTTERFLY

"Yes, my child. Every butterfly, before becoming the graceful creature which flies from flower to flower with magnificent wings, is an ugly caterpillar that creeps with effort. Thus the cabbage butterfly, which I have just shown you, is first a green caterpillar, which stays on the cabbages and gnaws the leaves. Jacques will tell you how much pains he takes to protect his cabbage patch from the voracious insect; for, you see, caterpillars have a terrible appetite. You will soon learn the reason.

"Most insects behave like caterpillars. On coming out of the egg, they have a provisional form that they must replace later by another. They are, as it were, born twice: first imperfect, dull, voracious, ugly; then perfect, agile, abstemious, and often of an admirable richness and elegance. Under its first form, the insect is a worm called by the general name of larva.

"You remember the lion of the plant-lice, the grub that eats the lice of the rosebush and, for weeks, without being able to satisfy itself, continues night and day its ferocious feasting. Well, this grub is a larva, that will change itself into a little lace-winged fly, the hemerobius, whose wings are of gauze and eyes of gold. Before becoming the pretty red ladybird with black spots, this pretty insect, which, in spite of its innocent air, crunches the plant-lice, is a very ugly worm, a slate-colored larva, covered with little points, and itself very fond of plant-lice. The June bug, the silly June bug, which, if its leg is held by a thread, awkwardly puffs out its wings, makes all preparations, and starts out to the tune of 'Fly, fly, fly!' is at first a white worm, a plump larva, fat as bacon, which lives under-ground, attacks the roots of plants, and destroys our crops. The big stag-beetle, whose head is armed with menacing mandibles shaped like the stag's horns, is at first a large worm that lives in old tree-trunks. It is the same with the capricorn, so peculiar for its long antennæ. And the worm found in our ripe cherries, which is so repugnant to us, what does it become? It becomes a beautiful fly, its wings adorned with four bands of black velvet. And so on with others.

"Well, this initial state of the insect, this worm, first form of youth, is called the larva. The wonderful change which transforms the larva into a perfect insect is called metamorphosis. Caterpillars are larvæ. By metamorphosis they turn into those beautiful butterflies whose wings, decorated with the richest colors, fill us with admiration. The Argus, now so beautiful with its celestial blue wings, was first a poor hairy caterpillar; the splendid swallow-tail began by being a green caterpillar with black stripes across it and red spots on its sides. Out of these despicable vermin metamorphosis has made those delightful creatures which only the flowers can rival in elegance.

"You all know the tale of Cinderella. The sisters have left for the ball, very proud, very smart. Cinderella, her heart full, is watching the kettle. The godmother arrives. 'Go,' says she, 'to the garden and get a pumpkin.' And behold, the scooped-out pumpkin changes under the godmother's wand, into a gilded carriage. 'Cinderella,' says she again, 'open the mouse-trap.' Six mice run out of it, and are no sooner touched by the magic wand than they turn into six beautiful dappled-gray horses. A bearded rat becomes a big coachman with a commanding mustache. Six lizards sleeping behind the watering-pot become green bedizened footmen, who immediately jump up behind the carriage. Finally the poor girl's shabby clothes are changed to gold and silver ones sprinkled with precious stones. Cinderella starts for the ball, in glass slippers. You, apparently, know the rest of it better than I.

"These powerful godmothers for whom it is play to change mice into horses, lizards into footmen, ugly clothes into sumptuous ones, these gracious fairies who astonish you with their fabulous prodigies, what are they, my dear children, in comparison with reality, the great fairy of the good God, who, out of a dirty worm, object of disgust, knows how to make a creature of ravishing beauty! He touches with his divine wand a miserable hairy caterpillar, an abject worm that slobbers in rotten wood, and the miracle is accomplished: the disgusting larva has turned into a beetle all shining with gold, a butterfly whose azure wings would have outshone Cinderella's fine toilette."

CHAPTER XXII

THE BIG EATERS

"INSECTS PROPAGATE themselves by eggs, which they lay, with admirable foresight, where the young will be sure to find nourishment. The little creature that comes from the egg is a larva, a feeble grub, which, most often, has to shift for itself, procure at its own risk food and shelter—the most difficult thing in this world. In these painful beginnings it cannot expect any help from its mother, dead some time before; for in insect life the parents generally die before the hatching of the eggs that produce the young. Without delay the little larva sets to work. It eats. It is its sole business, and a serious one, on which its future depends. It eats, not only to keep up its strength from day to day, but above all to acquire the plumpness necessary for its future metamorphosis. I must tell you—and this perhaps will surprise you—that an insect ceases to grow after attaining its final perfect form. It is known, too, that there are insects—among others, the butterfly of the silkworm—that do not take any nourishment at all.

"A cat is at first a tiny little pink-nosed creature, so small that it could rest in the hollow of the hand. In one or two months it is a pretty kitten that amuses itself at a mere nothing, and with its nimble paw whips the wisp of paper that one throws before it. Another year, and it is a tom-cat that patiently watches for mice or joins battle with its rivals on the roof. But, whether a tiny creature hardly able to open its little blue eyes, or a pretty playful kitten, or a big quarrelsome tom-cat, it has always the form of a cat.

"It is otherwise with insects. The swallow-tail, under its form of butterfly, is not first small, then medium, then large. When, for the first time, it opens its wings and takes flight, it is as large as it ever will be. When it comes out from under ground, where it lived as a grub, when for the first time it appears in the daylight, the June bug is such as you know it. There are little cats, but no little swallow-tails nor little June bugs. After the metamorphosis, an insect is what it will be to the end."

"But I have seen small June bugs flying round the willows in the evening," objected Jules.

"Those little June bugs are of a different kind. They will always remain the same. Never will they grow and become common June bugs, any more than a cat would grow into a tiger, which it resembles so much.

"The grub alone grows. At first very small on coming out of the egg, little by little it acquires a size in conformity with the future insect. It gathers the materials that the metamorphosis will use,—materials for the wings, antennæ, legs, and all those things that the larva does not have, but that the insect must have. Out of what will the big green worm that lives in dead wood, and must some day become a stag-beetle, make the enormous branched mandibles and the robust horny covering of the perfect insect? Of what will the larva make the long antennæ of the capricorn? Of what will the caterpillar make the large wings of the swallow-tail? Of that which the caterpillar, larva, and worm amass now, with thrifty hoarding of life-supporting matter.

"If the little pink-nosed cat were born without ears, paws, tail, fur, mustaches, if it were simply a little ball of flesh, and should some day have to acquire all at once, while asleep, ears, paws, tail, fur, mustaches, and many other things, is it not true that this work of life would necessitate materials gathered together beforehand and held in reserve in the fatty tissues of the animal? No thing can be made from noth-

GOAT MOTH

ing; the smallest hair of the cat's mustache shoots forth at the expense of the substance of the animal, substance which it acquires by eating.

"The larva is in precisely this case: it has nothing, or next to nothing, that the perfect insects must have. It must therefore amass, in view of future changes, materials for the change; it must eat for two: for itself first, and then for the insect that will come from its substance, transformed and, in a sense, recast. So the larvæ are endowed with an incomparable appetite. As I have said, to eat is their sole business. They eat night and day, often without stopping, without taking breath. To lose a mouthful, what imprudence! The future butterfly would perhaps have one scale less to its wings. So they eat gluttonously, take on a stomach, become big, fat, plump. It is the duty of larvæ.

"Some attack plants; they browse on the leaves, chew the flowers, bite the flesh of fruit. Others have a stomach strong enough to digest wood; they hollow out galleries in the tree-trunks, file off, grate, pulverize the hardest oak, as well as the tender willow. Others, again, prefer decomposed animal matter; they haunt infected corpses, fill their stomachs with rottenness. Still others seek excrement and feast on filth. They are all scavengers on whom has developed the high mission of cleansing the earth of its pollution. You would sicken at the mere thought of these worms that swarm in pus; yet one of the most important services, a providential service, is rendered by these disgusting eaters which clear away infection and give back its constituent elements to life. As if to make amends for its filthy needs, one of these larvæ will later be a magnificent fly, rivaling polished bronze in its brilliancy; another, a beetle perfumed with musk, its rich coat vying with gold and precious stones in splendor.

"But these larvæ devoted to the work of general sanitation cannot make us forget other eaters, of whom we are victims. The grubs of the June bug alone sometimes multiply so rapidly in the ground that immense tracts are denuded of vegetation, which is gnawed at the roots. The forester's shrubs, the farmer's harvests, the gardener's plants, just when everything seems prosperous, some fine morning, hang withered, smitten to death. The worm has passed that way, and all is lost. Fire could not have committed more frightful ravages. A miserable yellow louse, hardly visible, lives under ground, where it attacks the roots of the grape vine. It is called *phylloxera*. Its calamitous breed threatens to destroy all our vineyards. Some grubs, small enough to lodge in a grain of wheat, ravage the wheat in our granaries and leave only the bran. Others browse the lucerne so that the mower finds nothing left. Others, for years, gnaw at the heart of the wood of the oak, poplar, pine, and divers, other large trees. Others, which turn into those little white butterflies

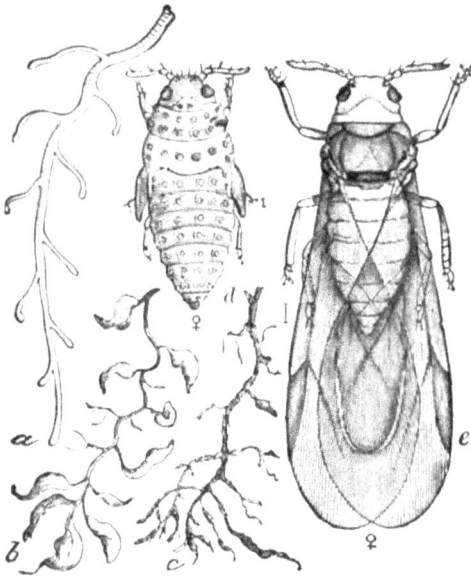

PHYLLOXERA

flying around the lamp in the evening and called moths, eat our cloth stuffs bit by bit, and finish by reducing them to rags. Others attack wainscoting, old furniture, and reduce them to powder. Others—But I should never get through if I were to tell you all. This little people to which we often disdain to pay the slightest attention, this little race of insects, is so powerful on account of the robust appetite of its larvæ, that man ought seriously to reckon with it. If a certain grub succeeds in multiplying beyond measure, whole provinces are threatened with the tragic fate of starvation. And we are left in perfect ignorance on the subject of these devourers! How can you defend yourself if the enemy is unknown to you? Ah, if I only had the management of these things! As for you, my dear children, while waiting for our talks to be resumed with more detail concerning these ravagers, remember this: the larvæ of insects are the great eaters of this world, the providential demolishers that finish the work of death and thus prepare for the work of life, since everything, or nearly everything passes through their stomach."

CHAPTER XXIII

SILK

"SOONER OR later, according to its species, a day comes when the larva feels itself strong enough to face the perils of metamorphosis. It has valiantly done its duty, since to stuff its paunch is the duty of a worm; it has eaten for two, itself and the matured insect. Now it is advisable to renounce feasting, retire from the world, and prepare itself a quiet shelter for the death-like sleep during which its second birth takes place. A thousand methods are employed for the preparation of this lodging.

"Certain larvæ simply bury themselves in the ground, others hollow out round niches with polished sides. There are some that make themselves a case out of dry leaves; there are others that know how to glue together a hollow ball out of grains of sand or rotten wood or loam. Those that live in tree-trunks stop up with plugs of sawdust both ends of the galleries they have hollowed out; those that live in wheat gnaw all the farinaceous part of the grain, scrupulously leaving untouched the outside, or bran, which is to serve them as cradle. Others, with less precaution, shelter themselves in some crack of the bark or of a wall, and fasten themselves there by a string which goes round their body. To this number belong the caterpillars of the cabbage butterfly and the swallow-tail. But especially in the making of the silk cell

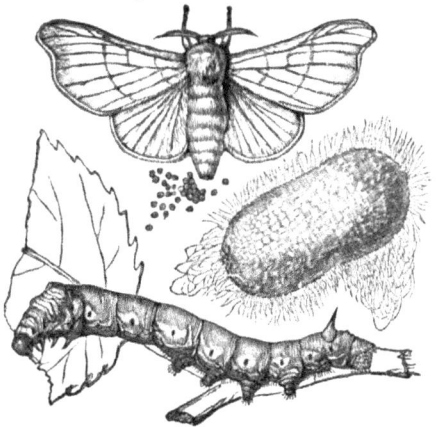

SILK WORM
EGGS, WORM, COCOON, AND BUTTERFLY

called cocoon is the highest skill of the larvæ shown.

"An ashy white caterpillar, the size of the little finger, is raised in large numbers for its cocoon, with which silk stuffs are made. It is called the silkworm. In very clean rooms are placed reed screens, on which they put mulberry leaves, and the young caterpillars come from eggs hatched in the house. The mulberry is a large tree cultivated on purpose to nourish these caterpillars; it has no value except for its leaves, the sole food of silkworms. Large tracts are devoted to its cultivation, so precious is the handiwork of the worm. The caterpillars eat the ration of leaves that is frequently renewed on the screens, and from time to time change their skin, according to their rate of growth. Their appetite is such that the clicking of their jaws is like the noise of a shower falling during a calm on the foliage of the trees. It is true that the room contains thousands and thousands of worms. The caterpillar gets its growth in four or five weeks. Then the screens are set with sprigs of heather, on which the worms climb when the time comes for them to spin their cocoons. They settle themselves one by one amid the sprigs and fasten here and there a multitude of very fine threads, so as to make a kind of network which will hold them suspended and serve them as scaffolding for the great work of the cocoon.

"The silk thread comes out of the under lip, through a hole called the spinneret. In the body of the caterpillar the silk material is a very thick, sticky liquid, resembling gum. In coming through the opening of the lip, this liquid is drawn out into a thread, which glues itself to the preceding threads and immediately hardens. The silk matter is not entirely contained in the mulberry leaf that the worm eats, any more than is milk in the grass that the cow browses. The caterpillar makes it out of the materials of its food, just as the cow makes milk of the constituents of her forage. Without the caterpillar's help man could never extract from the mulberry leaves the material for his costliest fabrics. Our most beautiful silk stuffs really take birth in the worm that drivels them into a thread.

"Let us return to the caterpillar suspended in the midst of its net. Now it is working at the cocoon. Its head is in continual motion. It advances, retires, ascends, descends, goes to right and left, while letting escape from its lip a tiny thread, which rolls itself loosely around the animal, sticks itself to the thread already in place, and finishes by forming a continuous envelope the size of a pigeon's egg. The silken structure is at first transparent enough to permit one to see the caterpillar at work; but as it grows thicker what passes within is soon hidden from view. What follows can easily be guessed. For three or four days the caterpillar continues to thicken the walls of the cocoon until it has exhausted its store of liquid silk. Here it is at last, retired from the world, isolated, tranquil, ready for the transfiguration so soon to take place. Its whole life, its long life of a month, it has worked in anticipation of the metamorphosis; it has crammed itself with mulberry leaves, has extenuated itself to make the silk for its cocoon, but thus it is going to become a butterfly. What a solemn moment for the caterpillar!

"Ah! my children, I had almost forgotten man's part in all this. Hardly is the work of the cocoon finished when he runs to the heather sprig, lays violent hands on the cocoons and sells them to the manufacturer. The latter, without delay, puts them into an oven and subjects them to the action of burning vapor to kill the future butterfly, whose tender flesh is beginning to form. If he delayed, the butterfly would pierce the cocoon, which, no longer capable of being unwound on account of its broken threads, would lose its value. This precaution taken, the rest is done at leisure. The cocoons are unwound in factories called spinning mills. They are put into a pan of boiling water to dissolve the gum which holds the successive windings together. A workwoman armed with a little heather broom stirs them in the water, in order to find and seize the end of the thread, which she puts on a revolving reel. Under the action of the machine the thread of silk unwinds while the cocoon jumps about in the hot water like a ball of wool when one pulls the yarn.

"In the center of the threadbare cocoon is the chrysalis, scorched and killed by the fire. Later the silk undergoes divers operations which give it more suppleness and luster; it passes into the dyer's vats where it takes any color desired; finally it is woven and converted into fabric."

CHAPTER XXIV

THE METAMORPHOSIS

"ONCE INCLOSED in its cocoon, the caterpillar withers and shrivels up, as if dying. First, the skin splits on the back; then, by repeated convulsions that pull it this way and that, the worm with much difficulty tears off its skin. With the skin comes everything: the case of the skull, jaws, eyes, legs, stomach and the rest. It is a general tearing-off. The ragged covering of the old body is finally pushed into a corner of the cocoon.

"What do they find then in the cells of silk? Another caterpillar, a butterfly? Neither. They find an almond shaped body, rounded at one end, pointed at the other, of a leathery appearance, and called a chrysalis. It is an intermediate state between the caterpillar and the butterfly. There can be seen certain projections which already indicate the shape of the future insect: at the large end can be distinguished the antennæ and the wings tightly folded crosswise on the chrysalis.

"The larvæ of the June bug, capricorn, stag-beetle, and other beetles pass through a similar state, but with more accentuated forms. The different parts of the head, wings, legs delicately folded at their sides, are very recognizable. But all is immobile, soft, white, or even transparent as crystal. This insect in outline is called a nymph. The name of chrysalis used for butterflies and that of nymph used for the other insects signify the same thing under somewhat different appearances. Both the chrysalis and the nymph are insects in process of formation—insects closely wrapped in swaddling-clothes, under which is finished the mysterious operation that will change their first structure from top to bottom.

"In a couple of weeks, if the temperature is favorable, the

chrysalis of the silkworm opens like a ripe fruit, and from its burst shell the butterfly escapes, all ragged, moist, scarcely able to stand on its trembling legs. Open air is necessary for it to gain strength, to spread and dry its wings. It must get out of the cocoon. But how? The caterpillar has made the cocoon so solid and the butterfly is so weak! Will it perish in its prison, the poor little thing? It would not be worth the trouble of going through so much to stifle miserably in the close cell, just as the end is attained!"

"Could it not tear the cocoon open with its teeth?" asked Emile.

"But, my innocent child, it has none, nor anything like them. It has only a proboscis, incapable of the slightest effort."

"With its claws then?" suggested Jules.

"Yes, if it had any strong enough. The trouble is, it is not provided with any."

"But it must be able to get out," persisted Jules.

"Doubtless it will get out. Has not every creature resources in the difficult moments of life! To break the hen's egg that imprisons it, the tiny little chicken has at the end of its beak a little hard point made on purpose; and the butterfly is to have nothing to open its cocoon? Oh, yes! But you would never guess the singular tool that it will use. It will use its eyes—"

"Its eyes?" interrupted Claire in amazement.

"Yes. Insects' eyes are covered with a cap of transparent horn, hard and cut in facets. A magnifying glass is needed in order to distinguish these facets, they are so fine; but, fine as they are, they have sharp bones which all together can, in time of need, be used as a grater. The butterfly begins then by moistening with a drop of saliva the point of the cocoon it wishes to attack, and then, applying an eye to the spot thus softened, it writhes, knocks, scratches, files. One by one the threads of silk succumb to the rasping. The hole is made, the butterfly comes out. What do you think about it? Do not animals sometimes have intelligence enough for four? Which of us would have thought of forcing the prison walls by striking them with the eye?"

"The butterfly must have studied a long time to think of that ingenious way?" queried Emile.

"The butterfly does not study, does not reflect; it knows at once what to do and how to do well whatever concerns it. Another has reflected for it."

"Who?"

"God himself! God, the great wise one. The silkworm butterfly is not pretty. It is whitish, tun-bellied, heavy. It does not fly like the others from flower to flower, for it takes no nourishment. As soon as it is out of the cocoon, it sets to work laying eggs; then it dies. Silkworm eggs are commonly called seed, a very good term, for the egg is the seed of the animal as the seed is the egg of the plant. Egg and seed correspond. They do not stifle all the cocoons in the vapor to wind them afterwards; they keep out a certain number so as to obtain butterflies and consequently eggs or seeds. These are the seeds which, the following year, produce the fresh brood of worms.

"All insects that are metamorphosed pass through the four states that I have just told you about: egg, larva, chrysalis or nymph, perfect insect. The perfect insect lays its eggs, and the series of transformations begins again."

CHAPTER XXV

SPIDERS

O NE MORNING, Mother Ambroisine was chopping herbs and cooked apples for a brood of little chickens hatched not long before. A large gray spider, letting itself slide the length of its thread, descended from the ceiling to the good woman's shoulders. At sight of the creature with long velvety legs, Mother Ambroisine could not suppress a cry of fear, and, shaking her shoulder, made the insect fall, and crushed it under her foot. "Spider in the morning stands for mourning," said she to herself. At this instant Uncle Paul and Claire entered.

"No, sir, it is not right," said Mother Ambroisine, "that we poor mortals should have so much useless trouble. Twelve little chickens are hatched out for us, bright as gold; and just as I am preparing them something to eat, a villainous spider falls on my shoulder."

And Mother Ambroisine pointed with her finger at the crushed insect with its legs still trembling.

"I do not see that those little chickens have anything to fear from the spider," remarked Uncle Paul.

SPIDER

"Oh! nothing, sir: the horrid creature is dead. But you know the proverb: 'Spider in the morning, mourning; spider at night, delight.' Everybody knows that a spider seen in the morning is a sign of bad luck. Our little chickens are in danger; the cats will claw them. You'll see, sir, you'll see."

Tears of emotion came to Mother Ambroisine's eyes.

"Put the little chickens in a safe place, watch the cats, and I will answer for the rest. The proverb of the spider is only a foolish prejudice," said Uncle Paul.

Mother Ambroisine did not utter another word. She knew that Maître Paul found a reason for everything, and on occasion was capable of pronouncing a eulogy on the spider. Claire, who saw this eulogy coming, ventured a question.

"I know: in your eyes all animals, however hideous they may be, have excellent excuses to plead: all merit consideration; all play a part ordained by Providence; all are interesting to observe and to study. You are the advocate of the good God's creatures; you would plead for the toad. But permit your niece to see there only an impulse of your kind heart, and not the real truth. What could you say in praise of the spider, horrid beast, which is poisonous and disfigures the ceiling with its webs?"

"What could I say? Much, my dear child, much. In the meantime, feed your little chickens and beware of cats if you want to prove the spider proverb false."

In the evening Mother Ambroisine, her large round spectacles on her nose, was knitting stockings. On her knees the cat slept and mingled its purring with the tick-tack of the needles. The children were waiting for the story of the spider. Their uncle began.

"Which of you three can tell me what spiders do with their webs, those fine webs stretched in the corners of the granary or between two shrubs in the garden!"

Emile spoke first. "It is their nest, Uncle, their house, their hiding-place."

"Hiding-place!" exclaimed Jules; "yes, I think it is more than that. One day I heard, between the lilac branches, a little shrill noise-he-e-e-e! A blue fly was entangled in a cobweb and trying to escape. It was the fly that was making the noise with its fluttering. A spider ran from the bottom of the silken funnel, seized the fly, and carried it off to its hole, doubtless to eat it. Since then I have thought spiders' webs were hunting nets."

"That is even so," said his uncle. "All spiders live on live prey;

they make continual war on flies, gnats, and other insects. If you fear mosquitoes, those insufferable little insects that sting us at night until they bring blood, you must bless the spider, for it does its best to rid us of them. To catch game, a net is necessary. Now, the net to catch flies in their flight is a cloth woven with silk, which the spider itself produces.

"In the body of the insect the silky matter is, as with caterpillars, a sticky liquid resembling glue or gum. As soon as it comes in contact with the air, this matter congeals, hardens, and becomes a thread on which water has no effect. When the spider wants to spin, the silk liquid flows from four nipples, called spinnerets, placed at the end of the stomach. These nipples are pierced at their extremity by a number of holes, like the sprinkler of a watering-pot. The number of these holes for all the nipples is roughly reckoned as a thousand. Each one lets its tiny little jet of liquid flow, which hardens and becomes thread; and from a thousand threads stuck together into one results the final thread employed by the spider. To designate something very fine there is no better term of comparison than the spider's thread. It is so delicate, in fact, that it can only just be seen. Our silk threads, those of the finest textures, are cables in comparison, cables of two, three, four strands, while this one, in its unequaled tenuity, contains a thousand. How many spiders' threads are required to make a strand of the thickness of a hair! Not far from ten. And how many elementary threads, such as issue from the separate holes of the spinneret! Ten thousand. To what a degree of tenuity then this silky matter can be reduced that stretches out in threads of which it takes ten thousand to equal the size of one hair! What marvels, my children, and only to catch a fly that is to serve for the spider's dinner!"

CHAPTER XXVI

THE EPEIRA'S BRIDGE

HERE UNCLE Paul caught Claire looking at him thoughtfully. It was evident that some change was taking place in her mind: the spider was no longer a repulsive creature, unworthy of our regard. Uncle Paul continued:

"With its legs, armed with sharp-toothed little claws like combs, the spider draws the thread from its spinnerets as it has need. If it wishes to descend, like the one this morning that came down from the ceiling on to Mother Ambroisine's shoulder, it glues the end of the thread to the point of departure and lets itself fall perpendicularly. The thread is drawn from the spinnerets by the weight of the spider, and the latter, softly suspended, descends to any depth it wishes, and as slowly as it pleases. In order to ascend again, it climbs up the thread by folding it gradually into a skein between its legs. For a second descent, the spider has only to let its skein of silk unwind little by little.

"To weave its web, each kind of spider has its own method of procedure, according to the kind of game it is going to hunt, the places it frequents, and according to its particular inclinations, tastes, and instincts. I will merely tell you a few words about the epeiræ, large spiders magnificently speckled with yellow, black, and silvery white. They are hunters of big game,—of green or blue damsel-flies that frequent the water-courses, of butterflies, and large flies. They stretch their web vertically between two trees and even from one bank of a stream to the other. Let us examine this last case.

"An epeira has found a good place for hunting: the dragon-flies, or blue and green damsel-flies, come and go from one tuft of reeds to another, sometimes going up, sometimes down the stream. Along its course are butterflies also, and horse-flies, or large flies that suck blood from cattle. The site is a good one. Now, then, to work! The epeira climbs to the top of a willow at the water's edge. There it matures its plan, an audacious one, the execution of which seems impossible. A suspension bridge, a cable which serves as support for the future web, must be stretched from one bank to the other. And observe, children, that the spider cannot cross the stream by swimming; it would perish by drowning if it ventured into the water. It must stretch its cable, its bridge, from the top of its branch without changing place. Never has an engineer found himself in such difficulties. What will the little creature do? Put your heads together, children; I am waiting for your ideas."

"Build a bridge from one side to the other, without crossing the water or moving away from its place? If the spider can do that it is cleverer than I am." Thus spoke Jules.

"Than I, too," chimed in his brother.

"If I did not already know," said Claire, "since you have just told us, that the spider does accomplish it, I should say that its bridge is impossible."

Mother Ambroisine said nothing, but by the slackening of the tick-tack of her needles, every one could see that she was much interested in the spider's bridge.

"Animals often have more intelligence than we," continued Uncle Paul; "the epeira will prove it to us. With its hind legs it draws a thread from its spinnerets. The thread lengthens and lengthens; it floats from the top of the branch. The spider draws out more and more; finally, it stops. Is the thread long enough? Is it too short? That is what must be looked after. If too long, it would be wasting the precious silky liquid; if too short, it would not fulfil the given conditions. A glance is thrown at the distance to be crossed, an exact glance, you may be sure. The thread is found too short. The spider lengthens it by drawing out a little more. Now all goes well: the thread has the wished-for length, and the work is done. The epeira waits at the top of its branch: the rest will be accomplished without help. From time to time it bears with its legs on the thread to see if it resists. Ah! it resists; the bridge is fixed! The spider crosses the stream on its suspension bridge! What has happened, then? This: The thread floated from the top of the willow. A breath of air blew the free end of the thread into the branches on the opposite bank. This end got entangled there; behold the mystery. The epeira has only to draw the thread to itself, to stretch it properly and make a suspension bridge of it."

"Oh, how simple!" cried Jules. "And yet not one of us would have thought of it."

"Yes, my friend, it is very simple, but at the same time very ingenious. It is thus with all work: simplicity in the means employed is a sign of excellence. To simplify is to have knowledge; to complicate is to be ignorant. The epeira, in its kind of construction, is science perfected."

"Where does it get that science, Uncle?" asked Claire. "Animals have not reason. Then who teaches the epeira to build its suspension bridges?"

"No one, my dear child; it is born with this knowledge. It has it by instinct, the infallible inspiration of the Father of all things, who creates in the least of His creatures, for their preservation, ways of acting before which our reason is often confounded. When the epeira, from the top of the willow, gets ready to spin its web, what inspires it with the audacious project of the bridge; what gives it patience to wait for the floating end of the thread to entwine in the branches of the other bank; what assures it of the success of a labor that it is performing perhaps for the first time, and has never seen done! It is the universal Reason that watches over creation, and takes among men the thrice-holy name of Providence."

Uncle Paul had won his case: in the eyes of all, even of Mother Ambroisine, spiders were no longer frightful creatures.

CHAPTER XXVII

THE SPIDER'S WEB

THE NEXT day the little chickens were all hatched and doing well. The hen had led them to the courtyard, and, scratching the soil and clucking, she dug up small seeds which the little ones came and took from their mother's beak. At the slightest approach of danger, the hen called the brood, and all ran to snuggle under her outspread wings. The boldest soon put their heads out, their pretty little yellow heads framed in their mother's black feathers. The alarm over, the hen began clucking and scratching again, and the little ones went trotting around her once more. Completely reassured, Mother Ambroisine forever renounced her proverb of the spider. In the evening Uncle Paul continued the story of the epeira.

"Since it must serve as a support to the silken network, the first thread stretched from one bank to the other must be of exceptional firmness. The epeira begins, therefore, by fixing both ends well; then, going and coming on the thread from one extremity to the other, always spinning, it doubles and trebles the strands and sticks them together in a common cable. A second similar cable is necessary, placed beneath the first in an almost parallel direction. It is between the two that the web must be spun.

"For this purpose, from one of the ends of the cable already constructed the epeira lets itself fall perpendicularly, hanging by the thread that escapes from its spinnerets. It reaches a lower branch, fastens the thread firmly to it, and ascends to the communicating bridge by the vertical thread it used for descending. The spider then reaches the other bank, still spinning, but without gluing this new strand of silk to the cable.

Arrived at the other side, it lets itself slide on to a branch conveniently placed, and there fastens the end of the thread that it has spun on its way from one bank to the other. This second chief piece of the framework becomes a cable by the addition of new threads. Finally the two parallel cables are made firm at each end by divers threads starting from it in every direction and attaching themselves to the branches. Other threads go out from this point and that, from one cable to the other, leaving between them, in the middle of the construction, a large open space, almost circular, destined for the net.

"Thus far the epeira has only constructed the framework of its building, a rough but solid framework; now begins the work of fine precision. The net must he spun. Across the open circular space that the divers threads of the framework leave between them, a first thread is stretched. The epeira stations itself right in the middle of this thread, central point of the web to be constructed. From this center numerous threads must start at equal distances from one another and be fastened to the circumference by the other end. They are called radiating lines. Accordingly the epeira glues a thread to the center and, ascending by the transverse thread already stretched, fixes the end of the line to the circumference. That done, it returns to the center by the line that it has just stretched; there it glues a second thread and immediately regains the circumference, where it fastens the end of the second line a short distance from the first one. Going thus alternately from the center to the circumference and from the circumference to the center by way of the last thread just stretched, the spider fills the circular space with radiating lines so regularly spaced that you would say they were traced with rule and compass by an expert hand.

"When the radiating lines are finished, the most delicate work of all is still left for the spider. Each of these lines must be bound by a thread that, starting at the circumference, twists and turns in a spiral line around the center, where it terminates. The epeira starts from the top of the web and, unwinding its thread, stretches it from one radiating line to another, keeping always at an equal distance from the outside thread. By thus circling about, always at the same distance from the preceding thread, the spider ends at the center of the radiating lines. The network is then finished.

"Now there must be arranged a little ambuscade from which the epeira can survey its web, a resting-room where it finds shelter from the coolness of the night and the heat of the day. In a little bunch of leaves close together the spider builds itself a silk den, a sort of funnel of close texture. That is its usual abiding place. If the weather is favorable and the passage of game abundant, morning and evening especially, the epeira leaves its den and posts itself, motionless, in the center of the web, to watch events more closely and run to the game quickly enough to prevent its escape. The spider is at its post, in the middle of the network, its eight legs spread out wide. It does not move, pretends to be dead. No hunter on the watch would have such patience. Let us copy its example and await the coming of the game."

The children were disappointed: at the moment when the story became the most interesting, Uncle Paul broke off his narrative.

"The epeira has interested me very much, Uncle," said Jules. "The bridge over the stream, the cobweb with its regular radiating lines, and the thread that twists and turns, getting nearer and nearer to the center, the room for ambush and rest-all that is very astonishing in a creature that does these wonderful things without having to learn how. Catching the game ought to be still more curious."

"Very curious indeed. Therefore, instead of telling you about the hunt, I prefer to show it to you. Yesterday, in crossing the field, I saw an epeira constructing its web between two trees on the little stream where such fine crayfish are caught. Let us get up early in the morning and go and see the chase."

CHAPTER XXVIII

THE CHASE

UNCLE PAUL had said: "Let us get up early in the morning." No one had to be called. One sleeps little when one is going to see an epeira hunt. About seven o'clock, with the sun shining bright, they were at the border of the stream. The

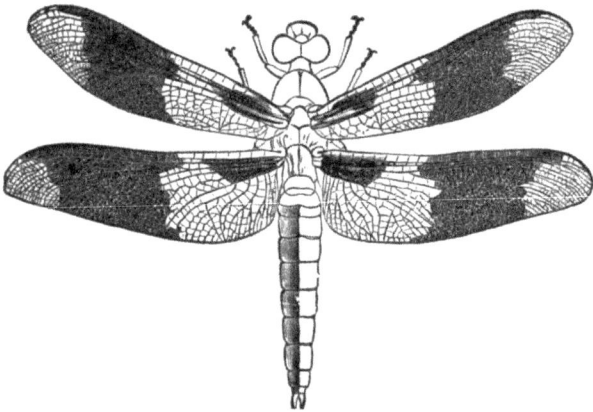

DAMSEL-FLY

cobweb was finished. Some dewdrops hanging to the threads shone like pearls. Hence the spider was not yet in the center of the net; no doubt it was waiting, before descending from its room, for the sun to dissipate the morning dampness. The party sat down on the grass for breakfast, at the very foot of the alder-tree to which were fastened the cables of the net. Blue damsel-flies flew from one tuft of rushes to another and chased each other playfully. Beware, you giddy ones, who will not know how to avoid the web by passing over and under it! Ah! it has happened; so much the worse for the victim. When one plays foolishly with one's companions, one must at least look where one is going. A dragon-fly is caught in the meshes of the web. With one wing free it struggles to escape. It shakes the web, but the cables hold in spite of the shaking. Threads in communication with the resting-room warn the epeira, by their agitation, of the important things taking place in the net. The spider hastily descends, but it does not get there in time. With a desperate stroke of its wing the dragon-fly frees itself and escapes, tearing a large hole in the web.

"Oh! how well it got out!" cried Jules. "A little more and the poor thing would have been eaten alive. Did you see, Emile, how quickly the spider ran down from its hiding place when it felt the web move? The hunt begins badly; the game escapes and the net is torn."

"Yes, but the spider is going to mend it," his uncle reassured him.

And, in fact, as soon as it had recovered from its misadventure, the epeira renewed the broken threads with delicate dexterity. The darning finished, the damage could hardly be detected. The spider now takes its place in the center of the network: the right moment for the chase has come, apparently, and it is advisable for it to pounce upon the game as quickly as possible, to avoid other misadventures. It spreads its eight feet in a circle, to receive the slightest movement that may come at any point of the web, and it waits, completely motionless.

The dragon-flies continue their evolutions. Not one is caught: the recent alarm has rendered them circumspect; they fly around the web to pass beyond it. Oh! oh! what is that coming so giddily and striking its head against the network? It is a little bumble-bee, all velvety and black, with a red stomach. It is caught. The epeira runs. But the captive is vigorous and formidable; perhaps it has a sting. The spider mistrusts it. It draws a thread from its spinneret and passes it quickly over the bee. A second silk string, a third, a fourth, soon subdue the captive's desperate efforts. Here is the bee strangled but still full of life, and menacing. To seize it in that state would be great imprudence: the epeira's life would be at stake. What must be done so as to leave nothing to fear from this dangerous prey? The spider possesses, folded under its head, two sharp-pointed fangs, which let flow a little drop of poison through a hole in their extremities. That is its hunting weapon. The epeira approaches cautiously, opens its fangs, stings the bee, and immediately moves aside. In the twinkling of an eye it is all over. The poison acts instantly: the bee trembles, its legs stiffen, it is dead. The spider carries it off to its silken chamber to suck it at leisure. When nothing but the skin is left, the spider will throw the remains of the bee far from its domicile, so as not to soil its web with a corpse that might frighten other game.

"It was done so quickly," complained Jules, "I did not see the spider's poisonous fangs. If we were to wait a little longer, another bumble-bee might perhaps come and then I should see it better."

"It is not necessary to wait," replied Uncle Paul. "If we proceed skilfully we can make the spider recommence its hunting manœuvers. All of you look attentively."

Uncle Paul searched among the field flowers for a moment and caught a large fly; then, holding it by one wing, put it near the web. The insect, beating about, gets entangled in the threads. The web shakes, the spider leaves its bee and runs, delighted with the fortunate chance that brings him prey again so quickly. The same manœuvers begin again. The fly is first strangled; the epeira opens its pointed fangs, stings the fly a little, and all is over. The victim trembles, stretches itself out, and ceases to move.

"Ah! that time I saw it," said Jules, satisfied at last.

"Claire, did you notice the fineness of the spider's fangs?" asked Emile. "I am sure that in your needle-case you haven't any such fine-pointed needles."

"I dare say not. As for me, what surprises me the most is not the fineness of the spider's fangs, but the quickness of the victim's death. It seems to me that a fly as large as this one ought not to die so quickly even from the coarser pricks of our needles."

"Very true," assented her uncle. "An insect transfixed by a pin still lives a long time; but if it is only pricked by the fine point of the spider's fangs, it dies almost instantly. But then, the spider takes care to poison its weapon. Its fangs are venomous; they are perforated by a minute canal through which the spider lets flow at will a scarcely visible little drop of liquid called venom, which the creature makes as it makes the silk liquid. The venom is held in reserve in a slender pocket placed in the interior of the fangs. When the spider pricks its prey, it makes a little of this liquid pass into the wound, and that suffices to bring speedy death to the wounded insect. The victim dies, not from the prick itself, but from the dreadful ravages wrought by the venom discharged into the wound."

Here Uncle Paul, in order to give his hearers a better view of the poisonous fangs, took the epeira with the tips of his fingers. Claire uttered a cry of fear, but her uncle soon calmed her.

"Don't be uneasy, my dear child: the poison that kills a fly will have no effect on Uncle Paul's hard skin."

And with the aid of a pin he opened the creature's fangs to show them in detail to the children, who were quite reassured.

"You must not be too frightened," he continued, "at the quick death of the fly and of the bumble-bee, and so look on spiders as creatures to be feared by us. The fangs of most of them would have great difficulty in piercing our skin. Courageous observers have let themselves be bitten by the various spiders of our country. The sting has never produced any serious results; nothing more than a redness less painful than that produced by the sting of a mosquito. At the same time, persons with a delicate skin ought to beware of the large kinds, were it only to spare themselves a passing pain. Without any excessive alarm we avoid the wasp's sting, which is very painful; let us avoid the spider's fangs in the same way without uttering loud cries at the sight of one of these creatures. We will resume the subject of the venomous insects. But it is late; let us go."

CHAPTER XXIX

VENOMOUS INSECTS

"YOU HAVE heard that certain creatures emit poison, that is to say, shoot from a distance into the face and on to the hands of those who approach a liquid capable of causing death, or at least of blinding or otherwise injuring them. Last week Jules found on the leaves of the potato-vines a large caterpillar armed with a curved horn."

"I know, I know," put in Jules. "It is the caterpillar, you told me, that turns into a magnificent butterfly called the sphinx Atropos. This butterfly, large as my hand, has on its back a white spot that frightens many people, for it has a vague resemblance to a death's-head. And besides, its eyes shine in the dark. You added that it was a harmless creature of which it would be unreasonable to be afraid."

"Jacques, who was weeding the potatoes," continued Uncle Paul, "knocked the sphinx caterpillar out of Jules's hands, and hastened to crush it with his big wooden shoe. 'What you are doing is very dangerous,' said the good Jacques. 'Handling poisonous creatures—of all things! Do you see that green venom? Don't get too close; the silly thing is not quite dead; it might yet throw some poison on you.' The worthy man took the green entrails of the crushed caterpillar for poison. Those entrails did not contain anything dangerous; they were green because they were swollen with the juice of the leaves that the poor thing had just eaten.

"Many persons are of the same opinion as Jacques: they are afraid of a caterpillar and the green of its entrails. They think that certain creatures poison everything they touch and throw out venom. Well, my dear children, you must bear this in mind, for it is a very important thing and frees us from foolish fears, while it puts us on guard against real danger: no animal of any kind, absolutely none, shoots venom and can harm us from a distance. To be convinced of this it suffices to know what venom really is. Divers creatures, large or small, are endowed with a poisoned weapon that serves them either as defense or to attack their prey. The bee is our best known venomous creature."

"What!" exclaimed Emile, "a bee is poisonous, the bee that makes honey for us?"

"Yes, the bee; the bee without which we could not have those honey cakes that Mother Ambroisine hands round when you are good. You don't think then of the stings that made you cry so?"

Emile blushed: his uncle had just revived unpleasant memories. From pure heedlessness he tried one day to see what the bees were doing. They say he even thrust a stick through the little door of the hive. The bees became incensed at this indiscretion. Three or four stung the poor boy on the cheeks and hands. He cried out most piteously, and thought himself done for. His uncle had much difficulty in consoling him. Compresses of cold water finally soothed his smarting pains.

"The bee is venomous," repeated Uncle Paul; "Emile could tell you that."

"The wasp too, then?" asked Jules. "One stung me once when I tried to drive it from a bunch of grapes. I did not say anything, but all the same I was not very comfortable. To think that such a tiny thing can hurt one so! It seemed as if my hands were on fire."

SOLITARY WASP AND NEST

"Certainly, the wasp is venomous; more so than the bee, in the sense that its sting causes greater pain. Bumble-bees are, too, as well as hornets, those large reddish wasps, an inch long, which sometimes come and gnaw the pears in the orchard. You must beware especially of hornets, my little friends. One sting from them, one only, would give you hours of horrible pain.

"All these insects have, for their defense, a poisoned weapon constructed in the same way. It is called the sting. It is a small,

AMERICAN HORNET

hard, and very pointed blade, a kind of dagger finer than the finest needle. The sting is placed at the end of the creature's stomach. When in repose, it is not seen; it is hidden in a scabbard that goes into its stomach. To defend itself, the insect draws it out of its sheath and plunges the point into the imprudent finger found within reach.

"Now it is not exactly the wound made by the sting that causes the smarting pain that you are familiar with. This wound is so slight, so minute, we cannot see it. We should hardly feel it were it made with a needle or a thorn as fine as the sting. But the sting communicates with a pocket of venom lodged in the creature's body, and, by means of a hollowed-out canal, it carries to the bottom of the wound a little drop of the formidable liquid. The sting is then drawn back. As to the venom, it stays in the wound and it is that, that alone, which causes those shooting pains that Emile could, if necessary, tell us about."

At this second attack from Uncle Paul, who dwelt on this misadventure in order to blame him for his heedless treatment of the bees, Emile blew his nose, although he did not need to. It was a way of hiding his confusion. His uncle did not appear to notice it, and continued:

"Scholars who have made a study of this curious question tell us of the following experiment, to make clear that it is really the venomous liquid introduced into the wound, and not the wound itself, that causes the pain. When one pricks oneself with a very fine needle, the hurt is very slight and soon passes off. I am sure Claire is not much frightened when she pricks her finger in sewing."

"Oh! no," said she. "That is so soon over, even if blood comes."

"Well, the prick of a needle, insignificant in itself, can cause sharp pains if the little wound is poisoned with the venom of the bee or wasp. The scholars I am telling you of dip the point of the needle into the bee's pocket of venom, and with this point thus wet with the venomous liquid give themselves a slight sting. The pain is now sharp and of long duration, more so than if the insect itself had stung the experimenter. This increase of pain is due to the fact that the comparatively large needle introduces into the wound more venom than could the bee's slender sting. You understand it now, I hope: it is the introduction of the venom into the wound that causes all the trouble."

"That is plain," said Jules. "But tell me, Uncle, why these scholars amuse themselves by pricking themselves with needles dipped in the bee's venom? It is a queer amusement, to hurt oneself for nothing."

"For nothing, Mr. Harum-scarum? Do you count as nothing what I have just told you? If I know it, must not others have taught me? Who are these others? They are the valiant investigators who learn about everything, observe and study everything, in order to alleviate our suffering. When they voluntarily prick themselves with poison, they propose to study in themselves, at their own risk and peril, the action of the venom, to teach us to combat its effects, which are sometimes so formidable. Let a viper or a scorpion sting us, and our life is in peril. Ah, then it is important to know exactly how the venom acts and what must be done to arrest its ravages; it is then that the scholars' researches are appreciated, researches that Jules looks upon as merely a queer amusement. Science, my little friend, has sacred enthusiasms that do not shrink from any test that may enlarge the sphere of our knowledge and diminish human suffering."

Jules, confused by his unfortunate remark, lowered his head and said not a word. Uncle Paul was on the point of getting vexed, but peace was soon restored and he continued the account of venomous creatures.

CHAPTER XXX

VENOM

"ALL VENOMOUS creatures act in the same way as the bee, wasp, and hornet. With a special weapon—needle, fang, sting, lancet—placed sometimes in one part of the body, sometimes in another, according to the species, they make a slight wound into which is instilled a drop of venom. The weapon has no other effect than that of opening a route for the venomous liquid, and this is what causes the injury. For the poison to act on us, it must come in contact with our blood by a wound which opens the way for it. But it has positively no effect on our skin, unless there is already a gash, a simple scratch, that permits it to penetrate into the flesh and mingle with the blood. The most terrible venom can be handled without any danger if the skin is not broken. Moreover, it can be put on the lips, on the tongue, even swallowed without any bad results. Placed on the lips, the hornet's venom produces no more effect than clear water; but if there is the slightest scratch the pain is atrocious. The viper's venom is equally harmless as long as it does not mingle with the blood. Courageous experimenters have tasted, swallowed it, and yet afterward were no worse off than before."

COPPER HEAD

"Is that true, Uncle? People have had the courage to swallow a viper's venom? Ah! I should not have been so brave." This from Claire.

"It is fortunate, my girl, that others have been so for us; and we ought to be very grateful to them, for by so doing they have taught us, as you will see, the most prompt and one of the most efficacious means to employ in case of accident."

"This viper's venom, which has no effect on the hand, lips, and tongue, is it much to be feared if it mingles with the blood?"

"It is terrible, my young lady, and I was just going to tell you about it. Let us suppose that some imprudent person disturbs the formidable reptile sleeping in the sun. Suddenly the creature uncoils itself in circles one above another, unwinds with the suddenness of a spring, and, with its jaws wide open, strikes you on the hand. It is done in the twinkling of an eye. With the same rapidity the viper refolds its spiral and draws back, continuing to menace you with its head in the center of the coil. You do not wait for a second attack, you flee; but, alas! the damage is done. On the wounded hand are seen two little red points, almost insignificant, mere needle pricks. It is not very alarming; you reassure yourself if you are in ignorance of what I so earnestly desire to teach you. Delusive innocuousness! See the red spots becoming encircled with a livid ring. With dull pains the hand swells, and the swelling extends gradually to the arm. Soon come cold sweats and nausea; respiration becomes painful, sight troubled, mind torpid, a general yellowness shows itself, accompanied by convulsions. If help does not arrive in time, death may come."

"You give us goose-flesh, Uncle," said Jules, with a shudder. "What should we poor things do if such a misfortune happened to us away from you, away from home? They say there are vipers in the underbrush of the neighboring hills."

"May God guard you from such a mischance, my poor children! But, if it befalls you, you must bind tight the finger, hand, arm, above the wounded part to prevent the diffusion of the venom in the blood; you must make the wound bleed by pressing round it; you must suck it hard to extract the venomous liquid. I told you venom has no effect on the skin. To suck it, therefore, is harmless if the mouth has no scratch. You can see that if, by hard suction and by pressure that makes the blood flow, you succeed in extracting all the venom from the wound, the wound itself is thenceforth of no importance. For greater surety, the wound should be cauterized as soon as possible with a corrosive liquid, aqua fortis or ammonia, or even with a red-hot iron. The effect of the cauterization is to destroy the venomous matter. It is painful, I acknowledge, but one must submit to it in order to avoid a worse evil. Cauterization is the doctor's business. The initial precautions, binding to prevent the diffusion of the venom, pressure to make the poisoned blood flow, hard suction to extract the venomous liquid, concern us personally, and all that must be done instantly. The longer it is put off, the more aggravated the evil. When these precautions are taken soon enough, it is seldom that the viper's bite has injurious consequences."

"You reassure me, Uncle. Those precautions are not difficult to take, if one does not lose one's presence of mind."

"Therefore it is important that we should all acquire the habit of using our reason in time of danger, and not let ourselves be overcome by ill-regulated fears. Man master of himself is half-master of danger."

CHAPTER XXXI

THE VIPER AND THE SCORPION

"**Y**OU JUST said," interposed Emile, "the bite of the viper, and not the sting. Then serpents bite, and do not sting. I thought it was just the other way. I have always heard they had a sting. Last Thursday lame Louis, who is not afraid of anything, caught a serpent in a hole of the old wall. He had two comrades with him. They bound the creature round the neck with a rush. I was passing, and they called me. The serpent was darting from its mouth something black, pointed, flexible, which came and went rapidly. I thought it was the sting and was much afraid of it. Louis laughed. He said what I took for a sting was the serpent's tongue; and to prove it to me, he put his hand near it."

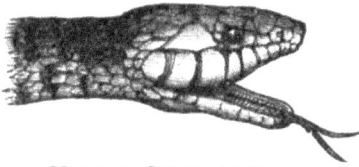

HEAD OF SNAKE SHOW-
ING FORKED TONGUE

"Louis was right," replied Uncle Paul. "All serpents dart a very flexible, forked, black filament between their lips with great swiftness. For many purposes it is the reptile's weapon, or dart; but in reality this filament is nothing but the tongue, a quite inoffensive tongue, which the creature uses to catch insects and to express in its peculiar manner the passions that agitate it by darting it quickly from between the lips. All serpents, without any exception, have one; but in our countries the viper alone possesses the terrible venomous apparatus.

"This apparatus is composed, first, of two hooks, or teeth, long and pointed, placed in the upper jaw. At the will of the creature they stand up erect for the attack or lie down in a groove of the gum, and hold themselves there as inoffensive as a stiletto in its sheath. In that way the reptile runs no danger of wounding itself. These fangs are hollow and pierced toward the point by a small opening through which the venom is injected into the wound. Finally, at the base of each fang is a little pocket full of venomous liquid. It is an innocent-looking humor, odorless, tasteless; one would almost think it was water. When the viper strikes with its fangs, the venomous pocket drives a drop of its contents into the canal of the tooth, and the terrible liquid is instilled into the wound.

"By preference the viper inhabits warm and rocky hills; it keeps under stones and thickets of brush. It is brown or reddish in color. On the back it has a somber zigzag band, and on each side a row of spots. Its stomach is slate-gray. Its head is a little triangular, larger than the neck, obtuse and as if cut off in front. The viper is timid and fearful; it attacks man only in self-defense. Its movements are brusk, irregular, and sluggish.

"The other serpents of our countries, serpents designated by the general name of snakes, have not the venomous fangs of the viper. Their bite therefore is not of importance, and the repugnance they inspire in us is really groundless.

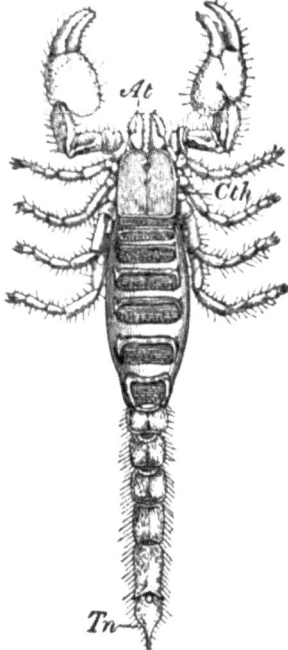
SCORPION SEEN FROM ABOVE

"Next to the viper there is in France no venomous creature more to be feared than the scorpion. It is very ugly and walks on eight feet. In front it has two pincers like those of the crayfish, and behind a knotty, curled tail ending in a sting. The pincers are inoffensive, despite their menacing aspect; it is the sting with which the end of the tail is armed that is venomous. The scorpion makes use of it in self-defense and to kill the insects on which it feeds. In the southern departments of France are found two different kinds of scorpions. One, of a greenish black, frequents dark and cool places and even establishes itself in houses. It leaves its retreat only at night. It can be seen then running on the damp and cracked walls, seeking wood-lice and spiders, its customary prey. The other, much larger, is pale yellow. It keeps under warm and sandy stones. The black scorpion's sting does not cause serious injury; that of the yellow may be mortal. When one of these creatures is irritated, a little drop of liquid can be seen forming into a pearl at the extremity of the sting, which is all ready to strike. It is the drop of venom that the scorpion injects into the wound.

"There are many other important things I could tell you about the venomous creatures of foreign countries, about divers serpents whose bite causes a dreadful death; but I hear Mother Ambroisine calling us to dinner. Let us go over rapidly what I have just told you. No creature, however ugly it may be, shoots venom or can do us any harm from a distance. All venomous species act in the same way: with a special weapon a slight wound is made; and into this wound a drop of venom is introduced. The wound, by itself, is nothing; it is the injected liquid that makes it painful and sometimes mortal. The venomous weapon serves the creature for hunting and for defense. It is placed in a part of the body that varies according to the species. Spiders have a double fang folded at the entrance of the mouth; bees, wasps, hornets, bumble-bees, have a sting at the end of the stomach and kept invisible in its sheath when in repose; the viper and all venomous serpents have two long hollowed-out teeth on the upper jaw; the scorpion carries a sting at the end of its tail."

"I am very sorry," said Jules, "that Jacques did not hear your account of venomous creatures; he would have understood that caterpillars' green entrails are not venom. I will tell him all these things; and if I find another beautiful sphinx caterpillar I will not crush it."

CHAPTER XXXII

THE NETTLE

AFTER DINNER, while their uncle read under the chestnut tree, the children scattered in the garden. Claire attended to her cuttings, Jules watered his vases, and Emile——Ah, giddy-pate, what should happen to him but another misfortune! A large butterfly was flying over the weeds that grow at the foot of the wall. Oh, what a magnificent butterfly! On the upper side its wings are red, fringed with black, with big blue eyes; underneath they are brown with wavy lines. It alights. Good. Emile makes himself small, approaches softly on tip-toe, puts out his hand, and, all at once, the butterfly is gone. But mark what follows. Emile draws his hand back quickly; it smarts, is red. The pain increases and becomes so bad that the poor boy runs to his uncle, his eyes swollen with tears.

"A venomous creature has stung me!" he cries. "See my hand, Uncle! It smarts—oh, how it smarts! Some viper has bitten me!"

At this word viper, Uncle Paul started. He rose and looked at the injured hand. A smile came to his lips.

"Impossible, my little friend; there is no viper in the garden. What foolishness have you been committing? Where have you been?"

"I ran after a butterfly, and when I put out my hand to catch it on the weeds at the foot of the wall, something stung me. See!"

NETTLE

"It is nothing, my poor Emile; go and dip your hand into the cool water of the fountain, and the pain will go away."

Quarter of an hour later they were talking of Emile's accident, he being quite recovered from his misadventure.

"Now that the pain is gone, does not Emile want to know what stung him?" asked his uncle.

"I certainly ought to know, so as not to be caught another time."

"Well, it is a plant called nettle. Its leaves, stems, slightest branches are covered with a multitude of bristles, stiff, hollow, and filled with a venomous liquid. When one of these bristles penetrates the skin, the point breaks, the little vial of venom opens and spills its contents into the wound. From that comes a smarting but not dangerous pain. You see, the nettle's bristles act like the weapons of venomous creatures. It is always a hollow point that makes a fine wound in the skin, and passes a drop of liquid into it, the cause of all the ill. The nettle is thus a venomous plant.

"I will also tell Emile that the beautiful butterfly for which he thoughtlessly thrust his hand into the tuft of nettles is called the Vanessa Io. Its caterpillar is velvety black with white spots. It also bristles with thorns. It does not make a cocoon. Its chrysalis, ornamented with bands that shine like gold, is suspended in the air by the end of its tail. The caterpillar lives on the nettle, of which it eats the leaves, notwithstanding their venomous bristles."

"In browsing on the venomous plant, how does the caterpillar manage so as not to poison itself?" Claire inquired.

"My dear child, you confound venomous with poisonous. Venomous is said of a substance that, introduced into the blood by any kind of a wound, causes injury in the manner of the viper's venom. Poisonous is said of a substance that, swallowed or introduced into the stomach, may cause death. Fatal drugs are poisonous: they kill if eaten or drunk. The liquid that flows from the viper's fangs and the scorpion's sting is venomous: it kills when it mixes with the blood; but it is not poisonous, for it can be swallowed with impunity. It is the same with the nettle's venom. So Mother Ambroisine gives the poultry chopped nettles, and the caterpillar of the Vanessa feeds without danger on the plant which, a little while ago, made Emile cry with pain. Of venomous plants we have in our country only nettles; but we have many poisonous plants that, when eaten, cause illness and even death. I must certainly tell you about them some day, so as to teach you to avoid them.

"The nettle's bristles remind me of the caterpillar's hairs. Many caterpillars have the skin quite bare. They are then perfectly inoffensive. They can be handled without any danger, however large they may be, even those that have a horn at the end of the back. They are no more to be feared than the silkworm. Others have bodies all bristly with hairs, sometimes very sharp and barbed, which can lodge in the skin, leave their points there, and thus produce lively itchings or even painful swellings. It is well then to mistrust velvety caterpillars, particularly those living in companies on oaks and pines, in large silk nests, and called processionary caterpillars. But here we have a word that calls for another story."

CHAPTER XXXIII

PROCESSIONARY CATERPILLARS

"WE FREQUENTLY see, at the ends of pine branches, voluminous bags of white silk intermixed with leaves. These bags are, generally, puffed out at the top and narrow at the bottom, pear-shaped. They are sometimes as large as a person's head. They are nests where live together a kind of very velvety caterpillars with red hairs. A family of caterpillars, coming from the eggs laid by one butterfly, construct a silk lodging in common. All take part in the work, all spin and weave in the general interest. The interior of the nest is divided by thin silk partitions into a number of compartments. At the large end, sometimes elsewhere, is seen a wide funnel-shaped opening; it is the large door for entering and departing. Other doors, smaller, are distributed here and there. The caterpillars pass the winter in their nest, well sheltered from bad weather. In summer they take refuge there at night and during the great heat.

"As soon as it is day, they set out to spread themselves on the pine and eat the leaves. After eating their fill they reënter their silk dwelling, sheltered from the heat of the sun. Now, when they are out on a campaign, be it on the tree that bears the nest, or on the ground passing from one pine to another, these caterpillars march in a singular fashion, which has given them the name of processionaries, because, in fact, they defile in a procession, one after the other, and in the finest order.

"One, the first come—for amongst them there is perfect equality—starts on the way and serves as head of the expedition. A second follows, without a space between; a third follows the second in the same way; and always thus, as many as there are caterpillars in the nest. The procession, numbering several hundreds, is now on the march. It defiles in one line, sometimes straight, sometimes winding, but always continuous, for each caterpillar that follows touches with its head the rear end of the preceding caterpillar. The procession describes on the ground a long and pleasing garland, which undulates to the right and left with unceasing variation. When several nests are near together and their processions happen to meet, the spectacle attains its highest interest. Then the different living garlands cross each other, get entangled and disentangled, knotted up and unknotted, forming the most capricious figures. The encounter does not lead to confusion. All the caterpillars of the same file march with a uniform and almost grave step; not one hastens to get before the others, not one remains behind, not one makes a mistake in the procession. Each one keeps its rank and scrupulously regulates its march by the one that precedes it. The file-leader of the troop directs the evolutions. When it turns to the right, all the caterpillars of the same line, one after the other, turn to the right; when it turns to the left, all, one after the other, turn to the left. If it stops, the whole procession stops, but not simultaneously; the second caterpillar first, then the third, fourth, fifth, and so on until the last. They would be called well-trained troops that, when defiling in order, stop at the word of command and close their ranks.

"The expedition, simply a promenade, or a journey in search of provisions, is now finished. They have gone far away from their nest. It is time to go home. How can they find it, through the grass and underbrush, and over all the obstacles of the road they have just traveled? Will they let themselves be guided by sight, obstructed though it be by every little tuft of grass; by the sense of smell, which wafted odors of every sort may put at fault? No, no; processionary caterpillars have for their guidance in traveling something better than sight or smell. They have instinct, which inspires them with infallible resources. Without taking account of what they do, they call to their service means that seem dictated by reason. Without doubt, they do not reason, but they obey the secret impulse of the eternal Reason, in whom and through whom all live.

"Now, this is what the processionary caterpillars do in order not to lose their way home again after a distant expedition. We pave our roads with crushed stone; caterpillars are more luxurious in their highways: they spread on their road a carpet of silk, they walk on nothing but silk. They spin continually on the journey and glue their silk all along the road. In fact, each caterpillar of the procession can be seen lowering and raising its head alternately. In the first movement, the spinneret, situated in the lower lip, glues the thread to the road that the procession is following; in the second, the spinneret lets the thread run out while the caterpillar is taking several steps. Then the head is lowered and lifted again, and a second length of thread is put in place. Each caterpillar that follows walks on the threads left by the preceding ones and adds its own thread to the silk, so that in all its length the road passed over is carpeted with a silky ribbon. It is by following this ribbon conductor that the processionaries get back to their home without ever losing their way, however tortuous the road may be.

"If one wishes to embarrass the procession, it suffices to pass the finger over the track so as to cut the silk road. The procession stops before the cut with every indication of fear and mistrust. Shall they go on! Shall they not go on! The heads rise and fall in anxious quest of the conductor threads. At last, one caterpillar bolder than the others, or perhaps more impatient, crosses the bad place and stretches its thread from one end of the cut to the other. A second, without hesitating, passes over on the thread left by the first, and in passing adds its own thread to the bridge. The others in turn all do the same. Soon the broken road is repaired and the defile of the procession continues.

"The processionary caterpillar of the oak marches in another way. It is covered with white hairs turned back and very long. One nest contains from seven to eight hundred individuals. When an expedition is decided on, a caterpillar leaves the nest and pauses at a certain distance to give the others time to arrange themselves in rank and file and form a battalion. This first caterpillar has to start the march. Following it, others place themselves, not one after another, like the processionaries of the pine, but in rows of two, three, four, and more. The troop, completed, begins to move in obedience to the evolutions of its file-leader, which always marches alone at the head of the legion, while the other caterpillars advance several abreast, dressing their ranks in perfect order. The first ranks of the army corps are always arranged in wedge formation, because of the gradual increase in the number of caterpillars composing it; the remainder are more or less expanded in different places. There are sometimes rows of from fifteen to twenty caterpillars marching in step, like well-trained soldiers, so that the head of one is never beyond the head of another. Of course the troop carpets its road with silk as it marches, so as to find its way back to its nest.

"The processionaries, especially those of the oak, retire to their nests to slough their skins, and these nests finally become filled with a fine dust of broken hairs. When you touch these nests, the dust of the hairs sticks to your hands and face, and causes an inflammation that lasts several days if the skin is delicate. One has only to stand at the foot of an oak where the processionaries have established themselves, to receive the irritating dust blown by the wind, and to feel a smart itching."

"What a pity the processionaries have those detestable hairs!" Jules exclaimed. "If they hadn't—"

"If they hadn't, Jules would much like to see the caterpillars' procession. Never mind; after all, the danger is not so great. And then, if one had to scratch one's self a little, it would not be a serious matter. Besides, we will turn our attention to the processionary of the pine, less to be feared than that of the oak. At the warmest part of the day we will go and look for a caterpillars' nest in the pine wood; but Jules and I will go alone. It would be too hot for Emile and Claire."

CHAPTER XXXIV

THE STORM

A ND, IN fact, it was very hot when Uncle Paul and Jules started out. With a burning sun, they were sure to find the caterpillars in their silk bag, where they do not fail to take refuge to shelter themselves from a light that is too glaring for them; at an earlier or later hour, the nests might be empty, and the journey a fruitless one.

His heart full of the naïve joys proper to his age, his mind preoccupied by the caterpillars and their processions, Jules walked at a good pace, forgetting heat and fatigue. He had untied his cravat and thrown his blouse back on his shoulders. A holly stick, cut by his uncle from the hedge, served him as a third leg.

In the meantime the crickets chirped louder than usual; frogs croaked in the ponds; flies became teasing and persistent; sometimes a breath of air all at once blew along the road and raised a whirling column of dust. Jules did not notice these signs, but his uncle did, and from time to time looked up at the sky. Masses of reddish mist in the south seemed to give him some concern. "Perhaps we shall have rain," said he; "we must hurry."

About three o'clock they were at the pine wood. Uncle Paul cut a branch bearing a magnificent nest. He had guessed right: all the caterpillars had returned to their lodging, perhaps in prevision of bad weather. Then they sat in the shade of a group of pines, to rest a little before returning. Naturally they talked about caterpillars.

"The processionaries, you told me," said Jules, "leave their nests to scatter over the pines and eat the leaves. There are, in fact, a great many branches almost reduced to sticks of dry wood. Look at that pine I am pointing at; it is half stripped of leaves, as if fire had passed over it. I like the way the processionaries travel, but I can't help pitying those fine trees that wither under the miserable caterpillar's teeth."

"If the owner of these pines understood his interests better," returned Uncle Paul, "he would, in the winter, when the caterpillars are assembled in their silk bags, have the nests collected and burn them, in order to destroy the detestable breed that will gnaw the young shoots, browse the buds, and arrest the tree's development. The harm is much greater in our orchards. Various caterpillars live in companies on our fruit trees and spin nests in the same way as the processionaries. When summer comes, the starveling vermin scatter all over the trees, destroying leaves, buds, shoots. In a few hours the orchard is shorn and the crop is destroyed in its budding. So it is necessary to keep a careful lookout for caterpillar nests, remove them from the tree before spring, and burn them, so that nothing can escape; the future of the crop depends on it. It is fortunate that several kinds of creatures, little birds especially, come to our aid in this war to the death between man and the caterpillar; otherwise the worm, stronger than man on account of its infinite number, would ravage our crops. But we will talk of the little birds another time; the weather is threatening, we must go."

See how the reddish mist in the south, thicker and darker every moment, has become a large black cloud visibly invading the still clear part of the sky. Wind precedes it, bending the tops of the pines like a field of grain. There rises from the soil that odor of dust which the dry earth gives forth at the beginning of a storm.

"We must not think of starting now," cautioned Uncle Paul. "The storm is coming; it will be upon us in a few minutes. Let us hurry and find shelter."

Rain forms in the distance like a dim curtain extending clear across the sky. The sheet of water advances rapidly; it would beat the fastest racing horse. It is coming, it has come. Violent flashes of lightning furrow it, thunder roars in its depths.

At a clap of thunder heavier than the others Jules starts. "Let us stay here, Uncle," says the frightened child; "let us stay under this big bushy pine. It doesn't rain here under cover."

"No, my child," replies his uncle, who perceives that they are in the very heart of the storm; "let us get away from this dangerous tree."

And, taking Jules by the hand, he leads him hastily through the hail and rain. Beyond the wood Uncle Paul knows of an excavation hollowed out in the rock. They arrive there just as the storm breaks with all its force.

They had been there a quarter of an hour, silent before the solemn spectacle of the tempest, when a flash of fire, of dazzling brightness, rent the dark cloud in a zigzag line and struck a pine with a frightful detonation that had no reverberation or echo, but was so violent that one would have said the sky was falling. The fearful spectacle was over in the twinkling of an eye. Wild with terror, Jules had let himself fall on his knees, with clasped hands. He was crying and praying. His uncle's serenity was undisturbed.

"Take courage, my poor child," said Uncle Paul as soon as the first fright had passed. "Let us embrace each other and thank God for having kept us safe. We have just escaped a great danger; the thunderbolt struck the pine under which we were going to take shelter."

"Oh, what a scare I had, Uncle!" cried the boy. "I thought I should die of it. When you insisted on hurrying away in spite of the rain, did you know that the bolt would strike that tree?"

"No, my dear, I knew nothing about it, nor could any one know; only certain reasons made me fear the neighborhood of the big branching pine, and prudence dictated the search for a less dangerous shelter. If I yielded to my fears, if I listened to the voice of prudence, let us give thanks to God, who gave me presence of mind at that moment."

"You will tell me what made you avoid the dangerous shelter of the tree, will you not?"

"Very willingly; but when we are all together, so that each one may profit by it. No one ought to ignore the danger one runs in taking shelter under a tree during a storm."

In the meantime the rain-cloud with its lightnings and thunders had moved on into the distance. On one side, the sun was setting radiant; on the opposite side, in the wake of the storm, the rainbow bent its immense bright arch of all colors. Uncle Paul and Jules started on their way, without forgetting the famous caterpillars' nest which might have cost them so dear.

CHAPTER XXXV

ELECTRICITY

JULES GAVE a lengthy account of the day to his brother and sister. At the part relating to the thunderbolt Claire trembled like a leaf. "I should have died of fright," said she, "if I had seen the lightning strike the pine." After the deeper emotion came curiosity, and they all agreed to beg their uncle for a talk on the subject of thunder. And so the next day Jules, Emile, and Claire gathered around their Uncle Paul to hear him tell them all about it. Jules broached the subject.

"Now that I am no longer afraid, will you please tell us, Uncle, why we should not take refuge under trees during a storm? Emile, I am sure, would like to know."

"I should first of all like to know what thunder is," said Emile.

"I too," said Claire. "When we know a little what thunder is, it will be much easier to understand the danger from trees."

"Quite right," commented their uncle, approvingly. "First let us see whether any one of you knows anything about thunder."

"When I was very small," Emile volunteered, "I used to think it was produced by rolling a large ball of iron made of resounding metal over the vault of the sky. If the vault broke anywhere, the ball was dashed to the ground and the thunder fell. But I don't believe that now. I am too big."

"Too big—a little fellow not so high as the first button on my vest! Say rather that your little reasoning powers are awakening and that the simple explanation of the iron ball no longer satisfies them."

Then Claire spoke. "I am not satisfied either with the explanations I used to give myself a while ago. With me, thunder was a wagon heavily loaded with old iron. It rolled on top of a sonorous vault. Sometimes a spark would flash out from under the wheels, the same as from a horse's hoof when it strikes a stone: that was the lightning. The vault was slippery and bordered with precipices. If the wagon happened to tip over, the load of old iron would fall to the ground, crushing people, trees, and houses. I laughed yesterday at my explanation, but I am no farther advanced now: I still know nothing at all about thunder."

"Your two thunders, varying to suit your infant imaginations, are based on the same idea, the idea of a sonorous vault. Well, know once for all that the blue vault of the sky is only an appearance due to the air which envelops us, and which, owing to the thickness of the envelope, has a beautiful blue color. Around us there is no vault, only a thick layer of air; and beyond that there is nothing for a vast distance until you come to the region of the stars."

"We will give up the blue vault," said Jules. "Emile, Claire, and I are persuaded there isn't any. Please go on."

"Go on? Here is where difficulty begins. Do you know, my children, that your questions are sometimes very embarrassing? 'Go on' is soon said; and, filled with unbounded faith in your Uncle Paul's knowledge, you expect an answer which, you feel sure, will satisfy your curiosity. You must, however, understand that there are innumerable things beyond your intelligence, and before you can grasp them you must attain to riper reason. With age and study many things will become clear that now are dark to you. In this number is the cause of thunder. I am very willing to tell you something about it; but if you do not understand all that I say you must blame your own premature curiosity. It is a difficult subject for you, very difficult."

"Only tell us about it," Jules persisted; "we will listen attentively."

"So be it. Air is not visible, one cannot take hold of it; if it were always at rest you would not, perhaps, suspect its existence. But when a violent wind bends tall poplars and scatters the leaves in eddies, when it uproots trees and carries off the roofs of buildings, who can doubt the existence of air? For wind is only air streaming irresistibly from place to place. Air, so subtle, so invisible, so peaceful in repose, is therefore in very truth a material substance, even a very brutal one when in violent motion. That is to say, a substance can exist, although at times nothing betrays its presence. We do not see it or touch it, are not sensible of it, and yet it is there, all about us; we are surrounded by it, live in the midst of it.

"Well, there is something still more hidden than air, more invisible, more difficult to detect. It is everywhere, absolutely everywhere, even in us; but it keeps itself so quiet that until now you have never heard of it."

Emile, Claire, and Jules exchanged glances full of meaning, trying to guess what it could be that was found ever where and that they did not yet know of. They were a hundred leagues from guessing what their uncle meant.

"You might seek in vain by yourselves all day, all the year, perhaps all your life; you would not find it. The thing I am speaking of, you understand, is singularly well hidden; scholars had to make very delicate researches to learn anything about it. Let us make use of the means they have taught us to bring it to light."

Uncle Paul took from his desk a stick of sealing-wax and rubbed it rapidly over his cloth sleeve; then he put it near a small piece of paper. The children were all eyes. Behold, the paper flies up and sticks to the sealing-wax. The experiment is repeated several times. Each time the paper rises unaided, starts off, and fastens on to the stick.

"The piece of sealing-wax, which formerly did not attract the paper, now does. The rubbing on the cloth has, then, developed in it something that cannot be seen, for the stick has not changed in appearance; and this invisible thing is nevertheless very real, since it can lift up the paper, draw it to the wax, and hold it glued there. This thing is called electricity. You can easily produce it by rubbing on cloth either a piece of glass or a stick of sulphur, resin, or sealing-wax. All these substances, when rubbed, will acquire the property of drawing to themselves very light objects, like small pieces of straw, little bits of paper, or particles of dust. This evening the cat shall teach us more about it, if it will be good."

CHAPTER XXXVI

THE EXPERIMENT WITH THE CAT

T HE WIND blew cold and dry. The storm of the day before had brought it on. Uncle Paul took this pretext to have the kitchen stove lighted in spite of Mother Ambroisine's remarks, who cried out at the unseasonableness of making a fire.

"Light up the stove in summer!" said she; "did one ever see the like? No one but our master would have such a notion. We shall be roasted."

Uncle Paul let her talk; he had his own idea. They sat down at the table. After eating its supper the big cat, never too warm, settled itself on a chair by the side of the stove, and soon, with its back turned to the warm sheet-iron, began to purr with happiness. All was going as desired; Uncle Paul's projects were taking an excellent turn. There was some complaint of the heat, but he took no notice.

"Ah! do you think it is for you the stove is lighted?" said he to the children. "Undeceive yourselves, my little friends: it is for the cat, the cat alone. It is so chilly, poor thing; see how happy it is on its chair."

Emile was on the point of laughing at his uncle's kindly attentions to the tom-cat, but Claire, who suspected serious designs, nudged him with her elbow. Claire's suspicions were well founded. When they had finished supper they resumed the subject of thunder. Uncle Paul began:

"This morning I promised to show you, with the cat's help, some very curious things. The time has come for keeping my word, provided Puss is agreeable."

He took the cat, whose hair was burning hot, and put it on his knees. The children drew near.

"Jules, put out the lamp; we must be in the dark."

The lamp put out, Uncle Paul passed and repassed his hand over the tom-cat's back. Oh! oh! wonderful! the beast's hair is streaming with bright beads; little flashes of white light appear, crackle, and disappear as the hand rubs; you would have said that sparks of fireworks were bursting out from the fur. All looked on in wonder at the tom-cat's splendor.

"That puts the finishing touch! Here is our cat making fire!" cried Mother Ambroisine.

"Does that fire burn, Uncle?" asked Jules. "The cat does not cry out, and you stroke him without being afraid."

"Those sparks are not fire," replied Uncle Paul. "You all remember the stick of sealing-wax which, after being rubbed on cloth, attracts little pieces of straw and paper. I told you that electricity, aroused by friction, is what makes the paper draw to the wax. Well, in rubbing the cat's back with my hand I produce electricity, but in greater abundance, so much so that it becomes visible where it was at first invisible, and bursts forth in sparks."

"If it doesn't burn, let me try," pleaded Jules.

Jules passed his hand over the cat's fur. The bright beads and their cracklings began again still stronger. Emile and Claire did the same. Mother Ambroisine was afraid. The worthy woman perhaps saw some witchcraft in the bright sparkles from her cat. The cat was then let loose. Besides, the experiment was beginning to give annoyance, and if Uncle Paul had not held the animal fast perhaps it would have begun to scratch.

CHAPTER XXXVII

THE EXPERIMENT WITH PAPER

"SINCE THE cat threatens to get cross, we will have recourse to another way of producing electricity.

"You fold lengthwise a good sheet of ordinary paper; then take hold of the double strip by each end. Next, you heat it just to the scorching point over a stove or in front of a hot fire. The greater the heat, the more electricity will be developed. Finally, still holding the strip by the ends alone, you rub it quickly, as soon as it is hot, on a piece of woolen cloth previously warmed and stretched over the knee. It can be rubbed on the trousers if they are woolen. The friction must be rapid and lengthwise of the paper. After a short rubbing the band is quickly raised with one hand, with great care not to let the paper touch against anything; if it did the electricity would be dissipated. Then without delay you bring up the knuckles of your free hand, or, better, the end of a key, near to the middle of the strip of paper; and you will see a bright spark dart from the paper to the key with a slight crackling. To get another spark you must go through the same operations again, for at the approach of the finger or key the sheet of paper loses all its electricity.

"Instead of making a spark, you can hold the electrified sheet flat above little pieces of paper, straw, or feathers. These light bodies are attracted and repelled in turn; they come and go rapidly from the electrified strip to the object which serves them as support, and from this to the strip."

Adding example to precept, Uncle Paul took a sheet of paper, folded it in a strip to give it more resistance, warmed it, rubbed it on his knee, and finally made a spark fly from it on the approach of his finger-joint. The children were full of wonder at the lightning that sprang from the paper with a crackle. The cat's beads were more numerous, but less strong and brilliant.

They say that Mother Ambroisine had much trouble that evening in getting Jules to go to bed; for, once master of the process, he did not tire of warming and rubbing. His uncle's intervention was necessary to put an end to the electric experiments.

CHAPTER XXXVIII

FRANKLIN AND DE ROMAS

THE NEXT day Claire and her two brothers could talk of nothing but the experiments of the evening before. It was their subject of conversation the whole morning. The cat's beads of fire and the flashes from the paper had greatly impressed them; so their uncle, in order to profit by this awakening of their attention, resumed as soon as possible his instructive talk.

"I am sure you are all three asking yourselves why, before telling you about thunder, I rubbed sealing-wax, a strip of paper, and the cat's back. You shall know, but first of all listen to a little story.

"More than a century ago a magistrate of the little town of Nérac, named de Romas, devised the most momentous experiment ever registered in the annals of science. One day he was seen going out into the country in a storm, with an enormous paper kite and a ball of twine. Over two hundred persons, keenly interested, accompanied him. What in the world was that celebrated magistrate going to do. Forgetful of his grave functions, did he propose some diversion unworthy of him? Was it to witness a puerile kite-flying that these curious ones flocked from all points of the town? No, no; de Romas was about to realize the most audacious project that man's genius has ever conceived; his bold purpose was to evoke the thunderbolt from the very depths of the clouds, and to call down fire from heaven.

"The kite that was to draw the thunderbolt from the midst of the storm-clouds and bring it into the intrepid experimenter's view did not differ from those familiar to you; only the hemp cord had through its entire length a copper thread. The wind having risen, the paper contrivance was thrown into the air and attained a height of about two hundred meters. To the lower end of the cord was attached a silk string, and this string was made fast under the stoop of a house, to shelter it from the rain. A little tin cylinder was hung to the hempen cord at one point and in touch with the metallic thread running through the cord. Finally, de Romas was furnished with a similar cylinder that had at one end a long glass tube as handle. It was with this instrument or this exciter, held in his hand by the glass handle, that he was to make the fire dart from the clouds, conducted by the copper thread of the kite to the metallic cylinder at the end of this thread. The silk cord and the glass handle served to prevent the passage of the thunderbolt, either into the ground or into the exciter's arm; for these substances have the property of not giving passage to electricity unless it is too strong. Metals, on the contrary, let it circulate freely.

"Such was the simple arrangement of the apparatus invented by de Romas to verify his audacious prevision. What is to be expected from this child's plaything thrown into the air to meet the thunder? Does it not seem to you foolish to suppose that such a plaything can direct the thunderbolt and master it? The magistrate of Nérac must, however, by wise meditations on the nature of thunder, have acquired the certainty of success, to dare thus, before hundreds of witnesses, to undertake this attempt, the failure of which would cover him with confusion. The result of this terrible conflict between thought and thunder cannot be in doubt: thought, as always, when well directed, will gain the upper hand.

"Behold, now, the clouds, forerunners of the storm, are coming near the kite. De Romas moves the exciter toward the tin cylinder suspended at the end of the cord, and suddenly there is a flash of light. It is produced by a dazzling spark which darts upon the exciter, crackles, emits a flash of lightning, and immediately disappears."

"That is just what we got yesterday evening," observed Jules, "when we put the end of a key near the strip of warmed and rubbed paper; it is what the cat's back showed us when it was stroked with the hand."

"The very same thing," replied his uncle. "Thunder, beads of fire from the cat, sparks from paper—all are due to electricity. But let us return to de Romas. We see that there is electricity, the thunderbolt in miniature, in the kite's string. It is inoffensive yet, on account of its feeble quantity; so de Romas does not hesitate to draw it forth with his finger. Every time he brings his finger near the cylinder, he draws a spark like that received by the exciter. Emboldened by his example, the spectators draw near and evoke the electric explosion. They crowd around the wonderful cylinder that now contains the fire from heaven, called down by man's genius; each one wishes to call forth the lightning, and each wishes to see sparkle between his fingers the fulminant substance descended from the clouds. So they play with the thunder for half an hour with impunity, when all at once a violent spark reaches de Romas and almost knocks him over. The hour of peril has come. The storm is getting nearer, stronger, every moment; thick clouds hover over the kite.

"De Romas summons up all his firmness; he quickly makes the crowd draw back and remains alone at the side of his apparatus, in the middle of the circle of spectators, who are beginning to get frightened. Then, with the aid of the exciter, he elicits from the metallic cylinder first strong sparks, capable of throwing a person down under the violence of the commotion, then ribbons of fire that dart in serpentine lines and burst with a crash. These ribbons soon measure a length of two or three meters. Any one struck by one of them would certainly perish. De Romas, fearing from moment to moment some fatal accident, enlarges the circle of curious spectators and ceases the perilous provocation of electric fire. But, braving imminent death, he continues his perilous observations at close range, with the same coolness as if he were engaged in the most harmless experiment. Around him there is heard a roaring like the continuous blast of a forge; an odor of burning is in the air; the kite-string is covered with a luminous envelope and forms a ribbon of fire joining heaven to earth. Three long straws, lying by chance on the ground, start up, jump, spring toward the string, fall, spring up again, and for some minutes entertain the spectators with their disordered evolutions."

"Last evening," Claire remarked, "the down of the feathers

and the little pieces of paper jumped in the same way between the electrified sheet of paper and the table."

"That is quite natural," said Jules, "since Uncle has just told us that the rubbed sheet of paper takes to itself the very essence of thunder, only in a very small quantity."

"I am glad to see you grasp the close resemblance between thunder and the electricity that we produce by rubbing certain bodies. De Romas made his perilous experiment on purpose to prove that resemblance. I said perilous experiment; you will see, in fact, what danger the audacious experimenter ran. Three straws, I told you, were jumping from the string to the ground, and from the ground to the string, when all at once everybody turned pale with fright: there came a violent explosion and a thunderbolt fell, making a large hole in the ground and raising a cloud of dust."

"My goodness!" gasped Claire. "Was de Romas killed?"

"No, de Romas was safe and beaming with joy: his previsions were verified with a success that bordered on the prodigious: it was demonstrated that a thunderbolt can be brought from the clouds within reach of the observer; he had proved that electricity is the cause of thunder. That, my children, was no trivial result, fit only to satisfy our curiosity: the nature of thunder being ascertained, it became possible to secure protection from its ravages, as I will tell you in the story of the lightning-conductor."

"De Romas, who made these important experiments at the peril of his life, must have been loaded with honors and riches by his contemporaries," said Claire.

"Alas! my dear child," replied her uncle, "things do not commonly happen that way. Truth rarely finds any free spot in which to plant itself; it has to fight against prejudice and ignorance. The battle is sometimes so painful, that men of strong will succumb to the task. De Romas, wishing to repeat his experiment at Bordeaux, was stoned by the mob, who saw in him a dangerous man evoking thunder by his witchcraft. He was obliged to flee in haste, abandoning his apparatus.

"A short time before de Romas, in the United States of North America, Franklin made similar researches on the nature of thunder. Benjamin Franklin was the son of a poor soap-manufacturer.

He found at home merely the requisite means for learning to read, write, and cipher; and yet he became by his learning one of the most remarkable men of his time. One stormy day in 1752 he went into the country near Philadelphia, accompanied by his son, who carried a kite made of silk tied at the four corners to two little glass rods. A metal tail terminated the apparatus. The kite was thrown up toward a storm-cloud. At first nothing happened to confirm the learned American's previsions: the string gave no sign of electricity. Rain came on. The wet string let the electricity circulate more freely; and Franklin, without thinking of the danger he ran, and transported with joy at thus stealing its secret from the thunder, elicited with his finger a shower of sparks strong enough to set fire to spirits of wine."

CHAPTER XXXIX

THUNDER AND THE
LIGHTNING-ROD

"**B**Y THEIR clever researches, Franklin, de Romas, and many others have revealed to us the nature of lightning; they have taught us, in particular, that when its quantity is small, it leaps to meet one's finger in bright, crackling sparks, without danger to the experimenter, and that all bodies containing it attract neighboring light substances, just as the kite-string attracted the straws in the experiment made by de Romas, and just as sealing-wax and rubbed paper attract the down of feathers. In short, they taught us that electricity is the cause of thunder.

"Now there are two distinct kinds of electricity, which are present in equal quantities in all bodies. As long as they are united, nothing betrays their presence; it is as if they did not exist. But, once separated, they seek each other across all obstacles, attract each other, and rush toward each other with an explosion and a flash of light. Then all is in complete repose until these two electric principles are again separated. The two electricities, therefore, supplement and neutralize each other; that is to say, they form something invisible, inoffensive, inert, that is found everywhere and is called neutral electricity. To electrify a body is to decompose its neutral electricity, to disunite the two principles which, when mixed, remain inert, but, separated from each other, manifest their wonderful properties and their violent tendency to recombination. Rubbing is one way of effecting the separation of the two electric principles, but it is far from being the only one. Every radical change in the inmost nature of a body also causes a manifestation of the two electricities. So clouds, which are water changed into vapor by the sun's heat, are often found to be electrified.

"When two differently electrified clouds come near together, immediately their contrary electricities run toward each other to recombine, and with a loud report there is a burst of flame that throws a bright and sudden light. This light is lightning; this burst of flame is a thunderbolt; the noise of the explosion is thunder. Finally, the electric spark can dart from a cloud electrified in one way to a spot on the ground electrified in the other.

"Generally you know a thunderbolt only by the sudden illumination it produces and the crash of its explosion. To see the thunderbolt itself you must overcome an unwarranted fear and look attentively at the clouds, the center of the storm. From moment to moment you can see a dazzling streak of light, simple or ramified, and of very irregular sinuous shape. A glowing furnace, metals at white heat, have not its brilliancy; the sun alone furnishes a comparison worthy the sovereign splendor of the thunderbolt."

"I saw the thunderbolt," put in Jules, "when it struck the big pine the day of the storm. For a moment I was blinded by its brightness, as if I had looked the sun full in the face."

"The next storm," said Emile, "I will watch the sky to see the ribbon of fire, but on condition that uncle is there. I should not dare to alone; it is so terrible."

"I, too," added Claire, "will do my best to overcome my fear, if Uncle is only there."

"I will be there, my children," their uncle promised them, "if my presence reassures you, for it is a most imposing sight, that of a stormy sky set on fire by lightning and full of the rumbling of the thunder. And yet, when from the bosom of the clouds there comes the dazzling flash of the thunderbolt and the whole region echoes with the crash of the explosion, a foolish fear dominates you; admiration has no further place in your mind, and your terrified eyes close at the magnificence of the electrical phenomena of the atmosphere, proclaiming with so much eloquence the majesty of the works of God. From your heart, congealed with fear, there comes no outburst of gratitude, for you do not know that at this moment, in the flashes of lightning, the uproar of the shower, of the thunder, and of the unchained winds, a great providential act is being accomplished. Thunder, in fact, is far more the cause of life than of death. In spite of the terrible but rare accidents that it causes, obeying in that the inscrutable decrees of God, it is one of the most powerful means that Providence employs to render the atmosphere wholesome, to clear the air we breathe of the deadly exhalations engendered by decay. We burn straw and paper torches in our rooms to purify the air; with its immense sheets of flame the thunderbolt produces an analogous effect in the surrounding atmosphere. Each of those lightning flashes that make you start with fear is a pledge of general salubrity; each of those claps of thunder that freeze you with fear is a sign of the great work of purification that is operating in favor of life. And who does not know with what delight, after a storm, the breast fills itself with pure air, when the atmosphere, purified by the fires of the thunderbolt, gives new life to all that breathe it! Let us beware then of a foolish terror when it thunders, but lift up our thoughts to God, from whom the thunder and the lightning have received their salutary mission.

"The thunderbolt, like everything in this world, plays a part in accord with the general well-being; but, again, like everything else, it can, in fulfilling the hidden purposes of an all-seeing Providence, cause here and there a rare accident that makes us forget the immense service it renders us. Let us always remember that nothing happens without the permission of our heavenly Father. A reverent fear of God ought to exclude all other fear. Let us, then, calmly examine the danger that a thunderbolt exposes us to. Let us remember above all that a thunderbolt by preference strikes the most prominent points of ground, for it is there that the opposite electricity, attracted by that of the storm-cloud, is present in greatest abundance, ready to unite with that which attracts it."

"The two electricities seeking reunion do their utmost to meet," said Claire, to fix the facts in her mind. "That of the ground, in its effort to reach the cloud, gains the top of a tall tree; that of the cloud, on its side, is impelled downward toward the tree. Then comes the moment when the two electricities, still attracting each other but no longer having a road open for their peaceful reunion, rush together with a crash. Then the streak of fire can't help reaching the tree. Is that it, Uncle!"

"My dear child, I could not have put it better myself. That is why, in fact, high buildings, towers, steeples, tall trees, are the points most exposed to fire from heaven. In the open country it would be very imprudent, during a storm to seek refuge from rain under a tree, especially a tall and isolated one. If the thunderbolt is to fall in the neighborhood, it will preferably be upon that tree, which forms a high point where the electricity of the ground accumulates, to get as near as possible to that of the cloud attracting it. The sad and deplorable instances every year of persons struck by lightning are for the most part confined to the imprudent who seek shelter from the rain under a tall tree."

"If you had not known about these things, Uncle," Jules here remarked, "we should have been killed the day of the storm, when I wanted to get under the tall pine-tree."

"It is very doubtful whether the thunderbolt, in destroying the tree, would have spared us. It is impious boldness to expose one's self to peril without a motive, and then to throw upon Providence the task of extricating us from our perilous situation. Heaven will help him who helps himself. We helped ourselves by fleeing from the dangerous tree, and we arrived home safe. But to help oneself effectively requires knowledge; so, to impress these things well on your mind, I emphasize once more the danger that, in time of storm, lurks in high towers, steeples, lofty buildings, and, above all, in tall and isolated trees. As for other precautions that are commonly recommended, such as not to run, in order not to cause a violent displacement of the air, and to shut the doors and windows in order to prevent a draught, they are of no value whatever: the direction taken by the thunderbolt is in no way affected by the air-currents. Railway trains, which run at high speed and displace the air with so much violence, are not more exposed to lightning than objects at rest. Every-day experience is a proof of it."

"When it thunders," said Emile, "Mother Ambroisine hurries to shut all the doors and windows."

"Mother Ambroisine is like a great many others who believe they are safe as soon as they cease to see the peril. They shut themselves up so as not to hear the thunder nor see the lightning; but that does not in the least lessen the danger."

"Then there are no precautions to be taken!" asked Jules.

"In the usual circumstances, none, unless it be this precaution: to be of good heart and rely on the will of God.

"To protect tall buildings, more menaced than others, we use a lightning-conductor, a wonderful invention due to Franklin's genius. The lightning-conductor is composed of a rod of iron, long, strong, and pointed, fastened to the top of the building. From its base starts another rod, also of iron, which runs along the roofs and walls, where it is fastened with staples, and plunges into damp ground or, better still, into a deep well of water. If a thunderbolt falls, it strikes the lightning-conductor, which is the nearest object to the cloud as well as the best suited to the electric current on account of its metallic nature. Besides, its pointed form has much to do with its efficacy. The bolt that strikes the metal lightning-conductor follows it and is dissipated in the depths of the earth without causing any damage."

CHAPTER XL

EFFECTS OF THE THUNDERBOLT

"A THUNDERBOLT OVERTHROWS, breaks, and rends bodies that do not permit electricity to circulate freely. It shatters rocks and throws the fragments great distances; it unroofs our dwellings; it splits the trunks of trees and divides the wood into little shreds; it overthrows walls, or even wrenches them from their foundations. In penetrating the ground, it melts the sand on its way and makes irregular glass tubes. It reddens, melts, and vaporizes metallic substances that give free passage to the electric current, such as metal chains, the iron wire of bells, the gilding of frames. Its preference, in short, is for objects made of metal. There are instances of persons left uninjured while the lightning consumed the various metallic objects worn or carried by them, such as gold-lace, metal buttons, and coins. It sets fire to piles of combustible matter like bundles of straw and stacks of dried fodder.

"A feeble electric spark, like those I taught you how to get from paper, makes but the slightest perceptible impression on us. At the very most, we feel a little prick at the point of communication. But with the help of powerful apparatus at the disposal of science, the electric shock becomes painful and can be dangerous, or even mortal. When one is struck by a rather strong spark, one feels, particularly in the joints, a sudden shock that makes one tremble and feel weak in the knees. With a still stronger spark, the whole body is seized with a sudden shaking so violent that the joints seem to be severed and one is knocked down by the stroke. Science possesses appliances powerful enough to kill an ox with the electric shock.

"The thunderbolt, a spark incomparably stronger than that

of our electric machines, gives to men and animals an extremely violent shock; it throws them down, injures them, and even kills them instantly. Sometimes a person thus struck bears traces, more or less deep, of burning; sometimes not the slightest wound is to be seen. Death is not, therefore, as a rule, due to any wounds inflicted by the thunderbolt, but to the sudden and violent shock given to the body. Sometimes death is only apparent: the electric shock simply suspends the primary vital functions, circulation and respiration. This state, which would end in death if prolonged, we can combat by giving the person struck the same care bestowed upon the drowned; that is to say, by seeking to revive by friction the respiratory movement of the breast. At other times the electric shock more or less paralyzes some part of the body, or perhaps only produces a passing disorder which wears off of itself."

CHAPTER XLI

CLOUDS

To finish his talk on lightning, the next morning Uncle Paul told them about clouds. The occasion, moreover, was very favorable. In one part of the sky great white clouds like mountains of cotton were piled up. The eye was delighted with the soft outlines of that celestial wadding.

"You remember," he began, "all those fogs that on damp autumn and winter mornings cover the earth with a veil of gray smoke, hide the sun, and prevent our seeing a few steps in front of us!"

"Looking into the air, you could see something like fine dust of water floating," said Claire; and Jules added:

"We played hide and seek with Emile in that kind of damp smoke. We could not see each other a few steps away."

"Well," resumed Uncle Paul, "clouds and fog are the same thing; only fog spreads about us and shows for what it is, gray, damp, cold; while clouds keep more or less above us and take on, with distance, a rich appearance. There are some of dazzling whiteness, like those you see over there; others of a red color, or golden-hued, or like fire; still others of the color of ashes, and others that are black. The color changes, too, from moment to moment. At sunset you will see a cloud begin with being white, then turn scarlet, then shine like a pile of embers, or like a lake of melted gold, and finally become dull and turn gray or black, according as the sun's rays strike it less and less. All that is a matter of illumination by the sun. In reality, clouds, however splendid in appearance, are formed of a damp vapor like that of fog. We can assure ourselves of this by a near approach."

"People can then mount as high as the clouds, Uncle?" Emile

asked.

"Certainly. All one needs is a pair of legs stout enough to climb to the top of a mountain. Often then clouds are under one's feet."

"And you have seen clouds underneath you?"

"Sometimes."

"That must be a very beautiful sight."

"So beautiful that words cannot express it. But it is not exactly a pleasure if the clouds mount and envelop you. You can be very much embarrassed by the obscurity of the fog alone. You lose your way; you become confused, without suspecting any danger in the most dangerous places, at the risk of falling into some abyss; you lose sight of the guides, who alone know the way and could save you from a false step. No, all is not roses up among the clouds. You will perhaps learn that some day to your cost. Meanwhile let us transport ourselves in imagination to the top of a cloud-capped mountain. If circumstances are favorable, here is what we shall see:

"Above our heads the sky, perfectly clear, presents no unusual appearance; the sun shines there in all its brilliancy. Down there at our feet, almost in the plains, white clouds spread themselves out. The wind sweeps them before it and drives them toward the summit. There they are, rolling and mounting up the side of the mountain. One would think they were immense flocks of cotton pushed up the slope by some invisible hand. Now and then a ray of sunlight penetrates their depths and gives them the brilliancy of gold and fire. The beautiful clouds behind which the sun disappears at its setting are not richer. What brilliant tints, what soft suppleness! They mount higher and higher. Now they roll up like a shining white band around the top of the mountain, and hide the view of the plain from us. Only the point where we are projects above the cloud-curtain, like an islet above the sea. At last this point is invaded, we are in the bosom of the clouds. Warm tints, soft outlines, striking views—all have disappeared. It is now only a dark fog that saturates with moisture and makes us feel depressed. Ah, if some breath of wind would make haste and sweep away these disagreeable clouds!

"That, my little friends, is what one does not fail to wish when one is in the clouds, which, so beautiful at a distance, are nothing but gloomy fog when close at hand. The spectacle of the clouds should be seen from afar. When in our curiosity we wish to examine certain appearances too closely, we sometimes find them deceptive; but we also find that, under a secondary brilliancy, which serves to adorn the earth, they hide realities of the first importance. The marvels of the clouds are only an appearance, an illusion of light; but under this illusion are concealed the reservoirs of rain, source of the earth's fecundity. God, by whom the smallest details of creation have been ordered, willed that the most common but also most necessary substances should serve as an

CIRRUS

ornament to the earth in spite of their really humble aspect; and he clothes them with a prestige dependent on the distance from which we are to contemplate them. The gray vapor of the clouds gives us rain. That is its chief utility. The sun illuminates it, and that suffices to transform it into a celestial tapestry in which the astonished eye finds the splendor of purple, gold and fire. That is its ornamental function.

"The height maintained by clouds is very variable and is generally less than you might suppose. There are clouds that lazily trail along the ground; they are the fogs. There are others that cling to the sides of moderately high mountains, and still others that crown the summits. The region where they are commonly found is at a height varying from 500 to 1500 meters. In some rather rare instances they rise to nearly four leagues. Beyond that eternal serenity reigns; clouds never mount there, thunder never rumbles, and snow, hail, and rain never form.

"Those clouds are called 'cirrus' that look sometimes like light flocks of curly wool, sometimes like drawn-out-filaments of dazzling whiteness, sharply contrasting with the deep blue of the sky. They are the highest of all the clouds. They are often a league high. When cirrus clouds are small and rounded and closely grouped in large numbers, so as to look like the backs of a flock of sheep, the sky thus covered is said to be dappled.

It is usually a sign that the weather is going to change.

"The name 'cumulus' is given to those large white clouds with round outlines which pile up, during the heat of summer, like immense mountains of cotton-wool. Their appearance presages a storm."

CUMULUS

"Then the clouds we see over there next to the mountains," queried Jules, "are cumulus? They look like piles of cotton. Will they bring us a storm?"

"I think not. The wind is driving them in another direction. The storm always takes place in their neighborhood. There! Hear that!"

A sudden light had just flashed through the flocks of the cumulus. After rather a long wait the noise of the thunder reached them, but greatly weakened by distance. Questions came quickly from Jules's and Emile's lips: "Why does it rain over there, and not here? Why does the noise of the thunder come after the lightning? Why—"

"We are going to talk about all that," said Uncle Paul; "but first let us learn the other forms of clouds. 'Stratus' is applied to clouds disposed in irregular bands placed in tiers on the horizon at sunrise or sunset. They are clouds that, in the fading daylight, especially in autumn, take the glowing tints of melted metals and of flame. The red stratus of the morning are followed by rain or wind.

STRATUS

"Finally, we give the name 'nimbus' to a mass of dark clouds of a uniform gray, so crowded together that it is impossible to distinguish one cloud from another. These clouds generally dissolve into rain. Seen from a distance, they often look like broad stripes extending in a straight line from heaven to earth. They are trails of rain.

"Now Emile may ask his questions."

CHAPTER XLII

THE VELOCITY OF SOUND

"UNDER THAT big white cloud that you call cumulus," said Emile, "there is at this very moment a storm. We have just seen the lightning and heard the thunder. Here, on the contrary, the sky is blue. So it does not rain everywhere at the same time. When rain is falling in one country, it is fine in others. And yet, when it rains here the whole sky is covered with clouds."

"You need only put your hand over your eyes to hide the sky," his uncle explained. "A cloud much farther off, but also much larger, produces the same effect: it veils what is surrounding us and makes it all cloudy. But that is only in appearance; beyond the region covered by the cloud the sky may be serene and the weather magnificent. Under the cumulus where the thunder is growling now, it rains, you may be sure, and the sky looks black. To the people in that region the surroundings present only a rainy appearance, because they are wrapped in clouds; if they were to go elsewhere, beyond the clouds, they would find the sky as serene as we have it here."

"With a fast horse they could, then," suggested Emile, "get from under the clouds, leave the rain, and come into fine weather; as also they could leave the sunshine and get into the rain under the clouds."

"Sometimes that would be possible, but more often not, because clouds can cover large areas. Besides, they travel, they go from one country to another, with such speed that the best horseman could not follow them in their course. You have all seen the shadow of the clouds run over the ground when the wind blows. Hills, valleys, plains, water-courses, all are crossed in less than no time. The shadow of a cloud passes over you at the moment you reach the top of a hill. Before you have taken three steps to descend into the valley, the shadow, with giant strides, is mounting the opposite slopes. Who could flatter himself that he could follow the cloud and keep under its cover?

"If rain sometimes falls over great stretches of country, it is never general, absolutely. If it should rain at one time over a whole province, what is that compared with the earth? A clod compared with a large field. Chased by the wind, clouds run hither and thither in the vast spaces of the atmosphere. They travel, and on their way throw a shadow or precipitate rain. Where they pass there is rain; everywhere else, no. In the same place there can even be both rain and fine weather, according as one is below or above the clouds. You know that on a mountain-top the clouds are sometimes beneath one. The plain under the clouds may receive a hard shower, while on the summit the sun shines without a single drop of rain."

"All that is easily understood," said Jules. "It is my turn now, Uncle, to ask you a question. From the storm-cloud that we see from here, there first came a flash of lightning; then, after waiting some time, the sound of the thunder was heard. Why do not the sound and the lightning come together?"

"Two things tell us of the thunderbolt: light and noise. The light is the flash of the lightning, the noise is thunder. Likewise in the discharge of firearms there is the light produced by the ignition of the powder and the noise resulting therefrom. At the scene of the explosion light and noise are coincident; but for persons at a distance the light, which travels at an incomparably greater velocity, arrives before the sound, which moves more slowly. If you note the discharge of a gun a considerable distance away, you see first the flash and smoke of the explosion, and do not hear the report until some time after; the more distant the explosion, the longer the time. Light travels an immense distance in an exceedingly short time. The flash of the explosion, therefore, reaches the eye at the very instant of its occurrence. If the sound does not arrive until after, it is because it travels much less rapidly and, in order to cover a considerable distance, requires considerable time, which is easily measured.

"Suppose ten seconds pass between the flash of a cannon's discharge and the arrival of the sound. The distance is measured between the place where the explosion occurred and that where it was heard. It is found to be 3400 meters. Sound, therefore, moves through the air, in a single second, a distance of 340 meters. That is a good rate of speed, comparable with that of the cannon-ball, but nothing, after all, in comparison with the inconceivable velocity of light.

BELLS RINGING

"The unequal rapidity with which sound and light travel accounts for the following fact. From a distance a wood-cutter is seen chopping wood, or a mason cutting stone. We see the ax strike the wood, the mallet tap the stone, and some time after we hear the sound."

"One Sunday before church," interposed Jules, "I was watching from a distance the ringing of the bell. I saw the tongue strike and the sound did not come until later. Now I see the reason."

"If you count the number of seconds between the appearance of the flash and the instant the thunder begins to be heard, you can tell what distance you are from the storm-cloud."

"Is a second very long!" Emile asked.

"It is about the length of one beat of the pulse. All we have to do, then, is to count, one, two, three, four, etc., without haste, but not too slowly, to have about the number of seconds. Note the instant the flash lights up the stormy cumulus, and count slowly until you hear the thunder."

With watchful eye and attentive ear all began the observation. Finally a flash was seen. They counted, the uncle beating time. One—two—three—four—five—At twelve came the thunder, but so faint that they could only just hear it.

"It took twelve seconds for the sound of the thunder to reach us," said Uncle Paul. "From what distance does it come, if sound travels 340 meters a second?"

"You must multiply 340 by twelve," replied Claire.

"Well, Miss, do it."

Claire made the calculation. The result was 4080 meters.

"The flash of lightning was 4080 meters away; we are more than a league from the storm-cloud," said her uncle.

"How easy that is!" exclaimed Emile. "You count one, two, three, four, and without moving you know how far away the thunderbolt has just fallen."

"The longer the time between the flash and the noise, the farther away is the cloud. When the report comes at the same time as the flash, the explosion is quite near. Jules knows that well since the day of the storm in the pine woods."

"I have heard that there is no longer any danger after the lightning is seen," said Claire.

"A thunderbolt is as rapid as light. An electric explosion is, therefore, ended as soon as the flash appears, and all danger is then passed; for the thunder, however loud it may be, can do no harm."

CHAPTER XLIII

THE EXPERIMENT WITH THE BOTTLE OF COLD WATER

U NCLE PAUL had rightly said, the evening before, that clouds are nothing but fog floating high in the air instead of spreading over the earth; but he had not said what fogs are composed of and how formed. So the next day he continued his talk on clouds.

"When Mother Ambroisine hangs the clothes she has just washed on the line, what does she do it for? To dry the linen, to free it from the water with which it is saturated. Well, what becomes of this water, if you please?"

"It disappears, I know," answered Jules, "but I should find it very hard to tell what becomes of it."

"This water is dissipated in the air, where it dissolves and becomes as invisible as the air itself. When you wet a heap of dry sand, the water permeates it throughout and disappears. It is true that the sand then takes a different appearance: it was dry before, it is wet afterward. The sand drinks the water that comes into contact with it. Air does the same: it drinks the moisture from the linen and becomes damp itself; and it drinks it so completely that all—air and water—remain as invisible as if the air held no foreign substance. Vapor is the name given to water thus made invisible, or in some sort aërial, that is to say resembling the air; and the reduction of water to this new state is called evaporation. The moisture of the linen we wish to dry evaporates; the water is dissipated in the air and thus becomes invisible vapor, which spreads in every direction at the will of the wind. The warmer it is, the quicker and more abundant

136

the evaporation. Have you not noticed that a wet handkerchief dries very quickly in a hot sun, and loses its moisture only very slowly if the weather is cloudy and cold?"

"Mother Ambroisine is always very glad when she has a fine day for her washing," Claire remarked.

"Remember, too, what happens after watering the garden. When, at close of a very warm day, we have to give a drink to those poor plants dying of thirst, something like this happens: The pump runs at its utmost capacity; you all make haste with your watering-pots; one goes here, another there, carrying water to the suffering plants, seed-plots, and potted flowers. Soon the garden has drunk copiously. How fresh it is then, how the plants, wilted by the heat, regain vigor and straighten up again, as happy as ever! You could almost think you heard them whispering to one another and telling how glad they were to be watered. If it could only stay that way! But, bah! the next day the earth is dry once more and all has to be done over again. What has become of the last evening's water? It has evaporated, dissolved into the air; and now it is perhaps traveling far away, at a great height, until, turned into a scrap of cloud, it falls again in rain. When Jules tires himself working the pump to water the flowers, has he ever thought that the water drawn from the well and spread over the ground sooner or later is dissipated in the immensities of the air to play its modest part in the formation of clouds?"

"In watering my garden," answered Jules, "I did not think I was watering the air more than anything else. But I see now: air is the great drinker. Of the contents of a watering-pot the plants take perhaps a handful; the air drinks up the rest. And that is why we have to do it all over again every day."

"And if you exposed a plateful of water to the sun what would finally become of it?"

"I will answer that," Emile hastened to reply. "Little by little, the water would turn into invisible vapor and there would be nothing but the plate left."

"What takes place at the expense of a plate of water, and of the moisture of the soil or wet linen, takes place also, on a vast scale, over the entire surface of the earth. The air is in contact with damp soil, with innumerable sheets of water, lakes, marshes, streams, rivers, brooks, above all with the sea, the immense sea, which presents thrice as much surface as the dry land. The great drinker, as Jules calls it, the air, must therefore drink to satiety and everywhere and always contain moisture, sometimes more, sometimes less, according to the heat.

"The air that is around us now, that invisible air in which the eye distinguishes nothing, nevertheless contains water that can be made visible. The means is very simple; all that is necessary is to cool the air a little. When you squeeze a wet sponge with the hand, you make water ooze out of it. Cold acts on moist air very much as the pressure of the hand on the sponge: it causes the moisture to distil in the form of minute drops. If Claire will go to the pump and fill a bottle with very cold water, I will show you this curious experiment."

Claire went to the kitchen and came back with a bottle full of the coldest water possible. Her uncle took the bottle, wiped it well with his handkerchief so that no trace of moisture should remain on the outside, and put it on an equally well-wiped plate.

Now the bottle, at first perfectly clear, becomes covered with a kind of fog which tarnishes its transparency: then little drops appear, run down its sides, and fall into the plate. At the end of a quarter of an hour there was enough water accumulated in the plate to fill a thimble.

"The drops of water now running down the outside of the bottle," Uncle Paul explained, "do not come, it is very clear, from the inside, for glass cannot be pierced by water. They come from the surrounding air, which cools off on touching the bottle and lets its moisture distil. If the bottle were colder, if full of ice, the deposit of liquid drops would be more abundant."

"The bottle reminds me of something of the same kind," said Claire. "When you fill a perfectly clean glass with very cold water, the outside of the glass immediately tarnishes and looks as if badly washed."

"That again is the surrounding air depositing its moisture on the cold side of the glass."

"Is that invisible moisture contained in the air abundant?" asked Jules.

"The invisible vapor of the air is always a thing so subtle, so disseminated, that it would take enormous volumes to make a small quantity of water. During the heat of summer, when the air holds the most vapor, it takes 60,000 liters of moist air to furnish one liter of water."

"That is very little," was Jules's comment.

"It is a great deal if one thinks of the immense volume of the atmosphere," replied his uncle, and then added:

"The experiment of the bottle teaches us two things: first, there is always invisible vapor in the air; in the second place, this vapor becomes visible and changes into fog, then into drops of water, by cooling. This return of invisible vapor to visible vapor or fog, then to a state of water, is called condensation. Heat reduces water to invisible vapor, and cold condenses this vapor, that is to say brings it back to a liquid state or at least to the state of visible vapor or fog. We will have the rest this evening."

CHAPTER XLIV

RAIN

"THE EXPLANATIONS of this morning account for the formation of clouds. A continual evaporation takes place on the surface of the damp earth as well as on the surface of the different sheets of water, lakes, ponds, marshes, streams, and above all the sea. The vapors formed rise into the air and remain invisible as long as the heat is sufficient. But since heat diminishes as the height increases, there comes a time when the vapors can no longer be kept in complete solution, and they condense into a mass of visible vapor, into a fog or cloud.

"When, after a chill encountered in the upper strata of the atmosphere, the cloud-mist reaches a certain degree of condensation, little drops of water form and fall in rain. At first very small, they increase in volume on the way by the union of other similar little drops. Their size on reaching us is proportioned to the height from which they fell, but never exceeds the limits suitable to the part rain is intended to play. If too large, the rain-drops would fall heavily on the plants they are to water, and would lay them flat on the ground, dead. And what would happen if the condensation of vapor, instead of taking place gradually, should be sudden? There would no longer descend from heaven rain-drops, but heavy columns of water, which, in their fall, would strip the trees of their branches, crush the harvests, and make the roofs of our houses fall in. But, far from taking this devastating form, rain falls in drops as if passed through a sieve placed by design in its passage to divide it and weaken the shock. On rare occasions, it is true, rain does reach us under so strange a disguise as to strike the ignorant with terror. Who would not be frightened when it rains blood or sulphur?"

"What do you say, Uncle?" interrupted Emile; "rains blood or sulphur? For my part, I should be dreadfully afraid."

"I too," said Claire.

"Is that true?" Jules asked, in his turn.

"True. You know well I only tell you true stories. There are rains of blood and sulphur, at least in appearance. It is proved that showers have been seen of which the drops left on the walls, roads, leaves of the trees, and clothes of passers-by, are red spots like blood. At other times, with the rain, there has fallen from the sky a fine dust, of a beautiful yellow, resembling sulphur. Did it really rain blood, sulphur? No. This so-called rain of blood or sulphur, causing foolish alarms, is ordinary rain stained with various sorts of dust raised from the ground by the wind. In the spring when, in mountainous countries, immense forests of fir-trees are in blossom, every breath of wind carries clouds of a fine yellow dust contained in the little flowers of the fir-tree. You can see a similar dust in all flowers, especially the lily."

"It is that dust that daubs your nose yellow when you smell a lily too close," declared Jules.

"Exactly. It is called pollen. Well, in falling at a distance, sometimes alone, sometimes accompanied by rain, the pollen gathered up from the forests by a breath of wind causes the so-called sulphur-rain."

"Your rain of blood or sulphur isn't at all terrifying," Claire remarked.

"Of course not; and yet whole populations have their hearts frozen with fear at the inoffensive fall of a whirlwind of pollen or red dust. They believe themselves visited with plagues, precursors of the end of the world. Ignorance is a pitiful thing, my dear children, and knowledge is a fine thing, even if it only served to deliver us from stupid terrors."

"In future," said Jules, stoutly, "it can rain sulphur or blood; if any one is afraid, it will not be I."

"There can also fall from the sky, with or without rain, various mineral substances, such as sand, for example, or powdered chalk, or dust from the roads. There is even mention of showers of small animals, caterpillars, insects, and very young toads. The marvelous feature of these rains disappears if one considers that a violent blast of wind can carry with it all light substances encountered in its course, and can transport them long distances before letting them fall again.

"At other times a rain of insects is due to something else besides transportation by the wind. Some kinds of grasshoppers, for example, gather together in immense swarms to go to another district when nutriment fails them. The emigrating band flies, as at a given signal, and passes through the air in the form of a great cloud that intercepts the daylight. The migration continues for days at a time, so numerous is the host. Then the voracious swarm alights, like a living storm, on the vegetation of some distant province. In a few hours grass, leaves of trees, grain, prairies—everything is browsed. The soil, as if ravaged by fire, hasn't a blade of grass left. Sometimes the people of Algeria die of hunger. The grasshopper has devoured their harvests.

"Volcanoes cause cinder-showers. Volcanic ashes is the name given to the calcined dust thrown up to a great height by volcanoes at the moment of their eruption. These powdered substances form enormous clouds, which produce in the daytime a darkness like that of the darkest nights, and which, falling to earth at a greater or less distance, stifle animals and plants under their showers of dust."

CHAPTER XLV

VOLCANOES

"IT IS not late yet, Uncle," said Jules; "you ought to tell us about those terrible mountains, those volcanoes that the showers of ashes come from."

At the word "volcano," Emile, who was already asleep, rubbed his eyes and became all attention. He too wanted to hear the great story. As usual, their uncle yielded to their entreaties.

"A volcano is a mountain that throws up smoke, calcined dust, red-hot stones, and melted matter called lava. The summit is hollowed out in a great excavation having the shape of a funnel, sometimes several leagues in circumference. That is what we call the crater. The bottom of the crater communicates with a tortuous conduit or chimney too deep to estimate. The principal volcanoes of Europe are: Vesuvius, near Naples; Etna in Sicily; Hecla in Iceland. Most of the time a volcano is either in repose or throwing up a simple plume of smoke; but from time to time, with intervals that may be very long, the mountain grumbles, trembles, and vomits torrents of fiery substances. It is then said to be in eruption. To give you a general idea of the most remarkable phenomena attending volcanic eruption, I will choose Vesuvius, the best known of the European volcanoes.

"An eruption is generally announced beforehand by a column of smoke that fills the orifice of the crater and rises vertically, when the air is calm, to nearly a mile in height. At this elevation it spreads out in a sort of blanket that intercepts the sun's rays. Some days before the eruption the column of smoke sinks down on the volcano, covering it with a big black cloud. Then the earth begins to tremble around Vesuvius; rumbling detonations under the ground are heard, louder and louder each moment, soon exceeding in intensity the most violent claps of thunder. You would think you heard the cannonades of a numerous artillery detonating ceaselessly in the mountain's sides.

"All at once a sheaf of fire bursts from the crater to the height of 2000 or 3000 meters. The cloud that is floating over the volcano is illumined by the redness of the fire; the sky seems inflamed. Millions of sparks dart out like lightning to the top of the blazing sheaf, describe great arcs, leaving on their way dazzling trails, and fall in a shower of fire on the slopes of the volcano. These sparks, so small from a distance, are incandescent masses of stone, sometimes several meters in dimension, and of a sufficient momentum to crush the most solid buildings in their fall. What hand-made machine could throw such masses of rock to such heights? What all our efforts united could not do even once, the volcano does over and over again, as if in play. For whole weeks and months these red blocks are thrown up by Vesuvius, in numbers like the sparks of a display of fireworks."

"It is both terrible and beautiful," said Jules. "Oh! how I should like to see an eruption, but far off, of course."

"And the people who are on the mountain?" questioned Emile.

"They are careful not to go on the mountain at that time; they might lose their lives, suffocated by the smoke or crushed by the shower of red-hot stones.

"Meantime, from the depths of the mountain, through the volcanic chimney, ascends a flux of melted mineral substance, or lava, which pours out into the crater and forms a lake of fire as dazzling as the sun. Spectators who, from the plain, anxiously follow the progress of the eruption, are warned of the coming of the lava-flood by the brilliant illumination it throws on the volumes of smoke floating in the upper air. But the crater is full; then the ground suddenly shakes, bursts open with a noise of thunder, and through the crevasses as well as over the edges of the crater the lava flows in streams. The fiery current, formed of dazzling and paste-like matter similar to melted metal, advances slowly; the front of the lava-stream resembles a moving rampart on fire. One can flee before it, but everything stationary is lost. Trees blaze a moment on contact with the lava and sink down, reduced to charcoal; the thickest walls are calcined and fall over; the hardest rocks are vitrified, melted.

"The flow of lava comes to an end, sooner or later. Then subterranean vapors, freed from the enormous pressure of the fluid mass, escape with more violence than ever, carrying with them whirlwinds of fine dust that floats in sinister clouds and sinks down on the neighboring plain, or is even carried by the winds to a distance of hundreds of leagues. Finally, the terrible mountain calms down, and peace is restored for an indefinite time."

"If there are towns near the volcanoes, cannot those streams of fire reach them? Cannot those clouds of ashes bury them?" asked Jules.

"Unfortunately all that is possible and has happened. I will tell you about it to-morrow, for it is time to go to bed now."

CHAPTER XLVI

CATANIA

"Yesterday," Uncle Paul resumed, "Jules asked me if the lava-streams could not reach towns situated near volcanoes. The following story will answer his question. It is about an eruption of Mount Etna."

"Etna is that volcano in Sicily where the big chestnut tree of a hundred horses is?" asked Claire.

"Yes. I must tell you that two hundred years ago there occurred in Sicily one of the most terrible eruptions on record. During the night, after a furious storm, the earth began to tremble so violently that a great many houses fell. Trees swayed like reeds shaken by the wind; people, fleeing distracted into the country to avoid being crushed under the ruins of their buildings, lost their footing on the quaking ground, stumbled, and fell. At that moment Etna burst in a fissure four leagues long, and along this fissure rose a number of volcanic mouths, vomiting, amid the crash of frightful detonations, clouds of black smoke and calcined sand. Soon seven of these mouths united in an abyss that for four months did not cease thundering, glowing, and throwing up cinders and lava. The crater of Etna, at first quite at rest, as if its furnaces had no connection with the new volcanic mouths, woke up a few days after and threw to a prodigious height a column of flames and smoke; then the whole mountain shook, and all the crests that dominated its crater fell into the depths of the volcano. The next day four mountaineers dared to climb to the top of Etna. They found the crater very much enlarged by the falling-in of the day before: its orifice, which before had measured one league, now measured two.

"In the meantime, torrents of lava were pouring from all

the crevasses of the mountain down upon the plain, destroying houses, forests, and crops. Some leagues from the volcano, on the seacoast, lies Catania, a large town surrounded then by strong walls. Already the liquid fire had devoured several villages, when the stream reached the walls of Catania and spread over the country. There, as if to show its strength to the terrified Catanians, it tore a hill away and transported it some distance; it lifted in one mass a field planted with vines and let it float for some time, until the green was reduced to charcoal and disappeared. Finally, the fiery stream reached a wide and deep valley. The Catanians believed themselves saved: no doubt the volcano would exhaust its strength by the time it covered the vast basin which the lava had just entered. But what an error of judgment! In the short space of six hours, the valley was filled, and the lava, overflowing, advanced straight toward the town in a stream half a league wide and ten meters high. It would have been all over with Catania if, by the luckiest chance, another current, whose direction crossed the first, had not come and struck against the fiery flood and turned it from its course. The stream, thus turned, coasted the ramparts of the town within pistol-shot, and turned toward the sea."

"I was very much afraid for those poor Catanians," interposed Emile, "when you spoke of that wall of fire, high as a house, going straight toward the town."

"All is not over yet," his uncle proceeded. "The stream, I told you, was going toward the sea. There was, then, a formidable battle between the water and the fire. The lava presented a perpendicular front of 1500 meters in extent and a dozen meters high. At the touch of that burning wall, which continued plunging further and further into the waves, enormous masses of vapor rose with horrible hissings, darkened the sky with their thick clouds, and fell in a salt rain over all the region. In a few days the lava had made the limits of the shore recede three hundred meters.

"In spite of that, Catania was still menaced. The stream, swollen with new tributaries, grew from day to day and approached the town. From the top of the walls the inhabitants followed with terror the implacable progress of the scourge. The lava finally reached the ramparts. The fiery flood rose slowly, but it rose ceaselessly; from hour to hour it was found to have risen a little higher. It touched the top of the walls, whereupon, yielding to the pressure, they were overthrown for the length of forty meters, and the stream of fire penetrated the town."

"My goodness!" cried Claire. "Those poor people are lost?"

"No, not the people, for lava runs very slowly, on account of its sticky nature, and one can be warned in time; it was the town itself that ran the greatest risk. The quarters invaded by the lava were the highest; from there the current could spread everywhere. So Catania seemed destined to total destruction, when it was saved by the courage of some men who attempted to battle with the volcano. They bethought themselves to construct stone walls, which, placed across the route of the oncoming stream, would change its direction. This device partly succeeded, but the following was the most efficacious. Lava streams envelop themselves in a kind of solid sheath, embank themselves in a canal formed of blocks coagulated and welded together. Under this covering the melted matter preserves its fluidity and continues its ravaging course. They thought, then, that by breaking these natural dikes at a well-chosen spot, they would open to the lava a new route across country and would thus turn it from the town. Followed by a hundred alert and vigorous men, they attacked the stream, not far from the volcano, with blows of iron bars. The heat was so great that each worker could strike only two or three blows in succession, after which he withdrew to recover his breath. However, they managed to make a breach in the solid sheath, when, as they had foreseen, the lava flowed through this opening. Catania was saved, not without great loss, for already the lava flood had consumed, within the town walls, three hundred houses and some palaces and churches. Outside of Catania, this eruption, so sadly celebrated, covered from five to six square leagues with a bed of lava in some places thirteen meters thick, and destroyed the homes of twenty-seven thousand persons."

"Without those brave men who did not hesitate, at the risk of being burnt alive, to go and open a new passage for the stream of fire, Catania would certainly have been lost," remarked Jules.

"Catania would have been all burnt down, there is no doubt. To-day its calcined ruins would be buried under a bed of cold lava, and there would be nothing left but the name of the large town that had disappeared. Three or four stout-hearted men revive the courage of the terrified population; they hope that heaven will aid them in their devotion, and, ready to sacrifice their lives, they prevent the frightful disaster. Ah! may God give you grace, my dear child, to imitate them in the time of danger; for, you see, if man is great through his intelligence, he is still greater through his heart. In my old age, when I hear you spoken of, I shall be more gladdened by the good you may have done than by the knowledge you may have acquired. Knowledge, my little friend, is only a better means of aiding others. Remember that well, and when you are a man bear yourself in danger as did those of Catania. I ask it of you in return for my love and my stories."

Jules furtively wiped away a tear. His uncle perceived that he had sown his word in good ground.

CHAPTER XLVII

THE STORY OF PLINY

"To teach you what the cinders thrown up by a volcano can do, I am now going to tell you a very old story, just as it was transmitted to us by a celebrated writer of those old times. This writer is called Pliny. His writing is in Latin, the great language of those days.

"It was in the year 79 of our era. Contemporaries of our Savior were still living. Vesuvius was then a peaceful mountain. It was not terminated then, as to-day, by a smoking cone, but by a table-land slightly concave, the remains of an old filled-up crater where thin grasses and wild vines grew. Very fertile crops covered its sides; two populous towns, Herculaneum and Pompeii, lay stretched at its base.

"The old volcano, which seemed forever lulled, and whose last eruptions went back to times beyond the memory of man, suddenly awakened and began to smoke. On the 23rd of August, about one o'clock in the afternoon, an extraordinary cloud, sometimes white, sometimes black, was seen hovering over Vesuvius. Impelled violently by some subterranean force, it first rose straight up in the form of a tree-trunk; then, after attaining a great height, it sank down under its own weight and spread out over a wide area.

"Now, there was at that time at Messina, a seaport not far from Vesuvius, an uncle of the author who has handed down these things to us. He was called Pliny, like his nephew. He commanded the Roman fleet stationed at this port. He was a man of great courage, never retreating from any danger if he could gain new knowledge or render aid to others. Surprised at the singular cloud that hovered over Vesuvius, Pliny immediately set out with his fleet to go to the aid of the menaced coast towns and to observe the terrible cloud from a nearer point. The people at the foot of Vesuvius were fleeing in haste, wild with fear. He went to the side where all were in flight and where the peril appeared the greatest."

"Fine!" cried Jules. "Courage comes to you when you are with those who are not afraid. I love Pliny for hastening to the volcano to learn about the danger. I should like to have been there."

"Alas! my poor child, you would not have found it a picnic. Burning cinders mixed with calcined stones were falling on the vessels; the sea, lashed to fury, was rising from its bed; the shore, encumbered with debris from the mountain, was becoming inaccessible. There was nothing to do but retreat. The fleet came to land at Stabiæ, where the danger, still distant, but all the time approaching, had already caused consternation. In the meantime, from several points on Vesuvius great flames burst forth, their terrifying glare rendered more frightful by the darkness caused by the cloud of cinders. To reassure his companions Pliny told them that these flames came from some abandoned villages caught by the fire."

"He told them that to give them courage," Jules conjectured, "but he himself well knew the truth of the matter."

"He knew it well, he knew the danger was great; nevertheless, overcome by fatigue, he fell into a deep sleep. Now, while he slept, the cloud reached Stabiæ. Little by little the court leading to his apartment was filled with cinders, so that in a short time he would not have been able to get out. They woke him to prevent his being buried alive and to deliberate on what was to be done. The houses, shaken by continual shocks, seemed to be torn from their foundations; they swayed from side to side. Many fell. It was decided to put to sea again. A shower of stones was falling—small ones, it is true, and calcined by the fire. As a protection from them, the men covered their heads with pillows, and going through the most horrible darkness, hardly relieved by the light of the torches they carried, they made their way toward the shore. There Pliny sat on the ground a moment to rest, when violent flames, accompanied by a strong smell of sulphur, put everybody to flight. He rose and then instantly fell back dead. The emanations, cinders, and smoke from the volcano had suffocated him."

"Poor Pliny! To be stifled to death like that by the horrible mountain, and he so courageous!" lamented Jules.

"Whilst the uncle was dying at Stabiæ, the nephew, left at Messina with his mother, was witness of what he relates to us. 'The night after my uncle's departure,' he tells us, 'the earth began to tremble violently. My mother hastened in alarm to waken me. She found me getting up to go and waken her. As the house threatened to collapse, we sat outside in the court, not far from the sea. With the carelessness of youth—I was then eighteen—I began to read. A friend of my uncle's came along. Seeing my mother and me both of us seated, and me with a book in my hand, he blamed us for our confidence and induced us to look out for our safety. Although it was seven o'clock in the morning, we could hardly see, the air was so obscured. At times buildings were so shaken that their fall was imminent at any moment. We followed the example of the rest and left the town. We stopped some distance off in the country. The wagons that were brought away swayed continually with the shaking of the ground. Even with their wheels blocked with stones they could hardly be held in place. The sea flowed back on itself: driven from the shore by the earthquake shocks, it receded from the beach and left a multitude of fish dry on the sand. A horrible black cloud came toward us. On its flanks were serpentine lines of fire like immense flashes of lightning. Soon the cloud descends, covering earth and sea. Then my mother begs me to flee with all the speed of my youth, and not to expose myself to imminent death by adapting my pace to hers, weighed down as she was by years. She would die content if she knew I was out of danger.'"

"And Pliny left his old mother behind in order to get away the faster?" queried Jules.

"No, my child, he did what you would all have done. He remained, sustaining and encouraging her, resolved to save himself with her or else die with her."

"Good!" cried Jules. "The nephew was worthy of his uncle.

And then what happened?"

"Then it was frightful. Cinders began to fall; darkness descended, so intense that they could see nothing. There was general confusion, outcry, and moaning. Wild with terror, the people fled at random, knocking down and treading on those who were in their way. The greater part were convinced that that night was the last, the eternal night that was to swallow the world. Mothers went groping for their children, lost in the crowd or perhaps crushed under the feet of the fugitives; they called them with doleful cries to embrace them once more and then die. Pliny and his old mother had seated themselves apart from the crowd. From time to time they were obliged to get up and shake off the cinders which would soon have buried them. At last the cloud dispersed and daylight reappeared. The earth was unrecognizable; everything had disappeared under a thick shroud of calcined dust."

"And the houses, were they buried in the cinders?" asked Emile.

"At the foot of the mountain the dust thrown up by the volcano lay deeper than the height of the tallest houses, and whole towns had disappeared under the enormous bed of cinders. Amongst these were Herculaneum and Pompeii. The volcano buried them alive."

"With the inhabitants?" inquired Jules.

"With a small number, for most of them, like Pliny and his mother, had time to flee to Messina. To-day, after being buried eighteen centuries, Herculaneum and Pompeii are exhumed by the miner's pick, just as they were when caught by the cloud of volcanic cinders. Vineyards cover them where they are not yet cleared."

"These vineyards, then, are the roofs of houses!" said Emile.

"Higher than the roofs of houses. The traveler who visits the quarters not yet uncovered, but made accessible by means of wells dug for the purpose, descends under-ground to a great depth."

CHAPTER XLVIII

THE BOILING POT

A s their uncle finished speaking, the postman came with a letter. A friend advised Uncle Paul to go to town on pressing business, and he wished to take advantage of the occasion to give his nephews the diversion of a little journey. He had Jules and Emile dressed in their Sunday clothes, and they set out to wait for the train at the neighboring station. At the station Uncle Paul went up to a grating behind which was a very busy man, and through a wicket he handed him some money. In exchange the busy man gave him three pieces of cardboard. Uncle Paul presented these pieces of cardboard to a man who guarded the entrance to a room. The man looked and let them enter.

Here they are in what is called the waiting-room. Emile and Jules open their eyes wide and say nothing. Soon they hear steam hissing. The train arrives. At its head is the locomotive, which slackens its speed so as to stop a moment. Through the window of the waiting-room Jules sees the people passing. Something preoccupies him: he is trying to understand how the heavy machine moves, what turns its wheels, which seem to be pushed by an iron bar.

They enter the railway car, the steam hisses, the train starts, and they are off. After a moment, when full speed had been gained: "Uncle Paul," said Emile, "see how the trees run, dance, and whirl around!" His uncle made him a sign to be silent. He had two reasons for this: first, Emile had just made a foolish remark, and, secondly, his uncle did not choose to notice the giddy-pate's self-betrayal in public.

Besides, Uncle Paul is not very communicative when traveling; he prefers to maintain a discreet reserve and keep silence. There are people whom you have never seen before, and perhaps will never see again, who immediately become very familiar with their traveling companions. Rather than hold their tongues they would talk to themselves. Uncle Paul does not like such people; he considers them weak-minded.

By evening the three travelers had returned, all much pleased with their trip. Uncle Paul had brought to a favorable conclusion his business in town. Emile and Jules each came back with an idea. When they had done honor to the excellent supper Mother Ambroisine had prepared on purpose to wind up the holiday with a little treat, Jules was the first to impart his idea to his uncle.

"Of all that I saw to-day," he began, "what struck me most was the engine at the head of the train, the locomotive that draws the long string of cars. How do they make it move? I looked well, but could not find out. It looks as if it went by itself, like a great beast on the gallop."

"It does not go by itself," replied his uncle; "it is steam that puts it in motion. Let us, then, first learn what steam is and what its power.

"When water is put on the fire, it first gets hot, then begins to boil, sending off vapor, which is dissipated in the air. If the boiling continues some time, it ends with there being nothing in the pot; all the water has disappeared."

"That is what happened to Mother Ambroisine day before yesterday," put in Emile. "She was boiling some potatoes, and having neglected to look into the pot for some time, she found them without a drop of water, half burnt. She had to begin all over again. Mother Ambroisine was not pleased."

"By heat," continued Uncle Paul, "water becomes invisible, intangible, as subtle as air. That is what is called vapor."

"You told us that the moisture in the air, the cause of fogs and clouds, is also vapor." This from Claire.

"Yes, that is vapor, but vapor formed only by the heat of the sun. Now, you must know that the stronger the heat, the more abundant is the vapor. If you put a pot full of water on the fire, the burning heat of the grate sets free incomparably more vapor than the temperature of a hot summer sun could. As long as it escapes freely from the pot, the vapor thus formed has nothing remarkable about it; so your attention has never been arrested by the fumes of a boiling pot. But if the pot is covered, covered tight, so as not to leave the slightest opening, then the steam, which tends to expand to an enormous volume, is furious to get out of its prison; it pushes and thrusts in all directions to remove the obstacles that oppose its expansion. However solid it may be, the pot ends by bursting under the indomitable pushing of the imprisoned steam. That is what I am going to show you with a little bottle, and not with a pot, which would not shut tight enough and the cover of which could be easily pushed off by the steam. And besides, even if I had a suitable pot, I should take care not to use it, for it might blow the house up and kill us all."

Uncle Paul took a glass vial, put a finger's breadth of water into it, corked it tightly with a cork stopper, and then tied the cork with a piece of wire. The vial thus prepared was put on the ashes before the fire. Then he took Emile, Jules, and Claire, and drew them quickly into the garden, to see from a distance what would happen, without fear of being injured by the explosion. They waited a few minutes, then boom! They ran up and found the vial broken into a thousand pieces scattered here and there with extreme violence.

"The cause of the explosion and the bursting of the bottle was the steam, which, having no way of escape, accumulated and exerted against its prison walls a stronger and stronger pressure as the temperature rose. A time then came when the vial could no longer resist the pressure of the steam, and it burst to pieces. They call elastic force the pressure exerted by steam on the inside of pots that hold it prisoner. The greater the heat, the stronger the pressure. With heat enough it may acquire an irresistible power, capable of bursting, not only a glass bottle, but also the thickest, most solid pots of iron, bronze, or any other very resistant material. Is it necessary to say that under those conditions the explosion is terrific? The fragments of the pot are thrown with a violence comparable to that of a cannon-ball or a bursting bomb. Everything standing in the way is broken or knocked down. Powder does not produce more terrific results. What I have just shown you with the glass vial is also not without some danger. You can be blinded with this dangerous experiment, which it is well to see once under proper precautions, but which it would be imprudent for you to repeat. I forbid you all, understand, to heat water in a closed vial; it is a game that might cost you your eyesight. If you should disobey me on this point, good-by to stories; I would not keep you with me any longer."

"Don't be afraid, Uncle," Jules hastened to interpose; "we will be careful not to repeat the experiment; it is too dangerous."

"Now you know what makes the locomotive and a great many other machines move. In a strong boiler, tightly closed, steam is formed by the action of a hot furnace. This steam, of an enormous power, makes every effort to escape. It presses particularly on a piece placed for that purpose, which it chases before it. From that a movement results that sets everything going, as you will see in the case of the locomotive. To conclude, let us remember that in every steam engine the essential thing, the generator of the force, is a boiler, a closed pot that boils."

CHAPTER XLIX

THE LOCOMOTIVE

Uncle paul showed his nephews the following picture, and explained it to them.

"This picture represents a locomotive. The boiler where the steam is generated, the boiling pot, in short, forms the greater part of it. It is the large cylinder that goes from one end to the other, borne on six wheels. It is built of solid iron plates, perfectly joined together with large rivets. In front the boiler terminates in a smoke-stack; behind, in a furnace, the door of which is represented as open. A man, called a stoker, is constantly occupied in filling the furnace with pit-coal, which he throws in by the shovelful; for he must keep up a very hot fire to heat the volume of water contained in the boiler and obtain steam in sufficient quantity. With an iron bar he pokes the fire, arranges it, makes it burn fast. That is not all: skilful arrangements are made to utilize the heat and warm the water quickly. From the end of the furnace start numerous copper pipes which traverse the water from one end to the other of the boiler, and terminate at the smoke-stack. You will see some in B where the picture supposes a part of the casing taken away to show the interior. The flame of the furnace runs through these pipes, themselves surrounded by water. By this means the fire is made to circulate through the very midst of the water, and so steam is obtained very quickly.

"Now look at the front of the locomotive. In A is seen a short

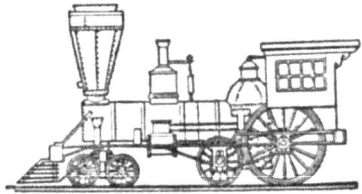

An old-time Locomotive

cylinder closed tightly, but represented in the picture with a part of the outside removed to show what is within. There are two of these cylinders, one on the right, the other on the left of the locomotive. Inside the cylinder is an iron stopper called a piston. The steam from the boiler enters the cylinder alternately

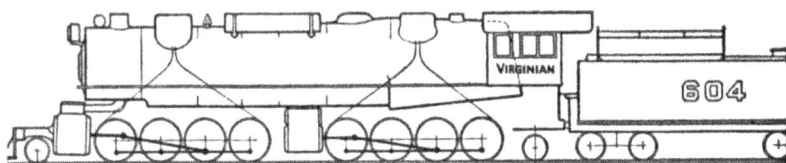

A MODERN LOCOMOTIVE

in front of and behind the piston. When the steam comes in front, what is behind escapes freely into the air by an orifice that opens of itself at the right moment. This escaping steam ceases to press on the piston, since it finds its prison open and that it can get out. We do not try to force doors when other outlets are open. So does steam act: the instant it can escape freely, it ceases to push. The entering steam, on the contrary, finds itself imprisoned. It pushes the piston, therefore, with all its strength and drives it to the other end of the cylinder. But then the rôles immediately change. The steam that hitherto has been pushing, escapes into the air and ceases to act, while on the other side a jet of steam rushes in from the boiler and begins to push in the contrary direction."

"Let me repeat it," said Jules, "to see if I have understood it properly. Steam comes from the boiler, where it forms unceasingly. It goes into the cylinder before and behind the piston by turns. When it gets in front, that behind escapes into the air and no longer pushes; when it gets behind, that in front escapes. The piston, pushed first one way, then the other, alternately, must advance and retreat, go and come, in the cylinder. And then?"

"The piston is in the form of a solid iron rod that enters the cylinder through a hole pierced in the middle of one of the ends, and just large enough to give free passage to the rod, without letting the steam escape. This rod is bound to another iron piece called a crank, and finally the crank is attached to the neighboring wheel. In the picture all these things can easily be seen. The piston, advancing and retreating in turn in the cylinder, pushes the crank forward and back, and the crank thus makes the great wheel turn. On the other side of the locomotive the same things are taking place by means of a second cylinder. Then the two great wheels turn at the same time and the locomotive moves forward."

"It isn't so hard as I thought," Jules remarked. "Steam pushes the piston, the piston pushes the crank, the crank pushes the wheel, and the engine moves."

"After acting on the piston, the steam enters the same chimney that the smoke comes out of. So you can see this smoke-stack sometimes throwing out white puffs, sometimes black. These latter are smoke coming from the furnace through the tubes that go through the water; the others come from the steam thrown out of the cylinders after each stroke of the piston. These white puffs, in rushing violently from the cylinder to the smoke-stack after acting on the piston, make the noise of the engine as it moves."

"I know: pouf! pouf! pouf!" exclaimed Emile.

"The locomotive carries with it a supply of coal to feed the fire, and a supply of water to renew the contents of the boiler as fast as evaporation may require. These supplies are carried in the tender; that is to say, in the vehicle that comes immediately behind the locomotive. On the tender are the stoker, who tends the furnace, and the engineer, who controls the passage of the steam into the cylinders."

"The man in the picture is the engineer?" Emile asked.

"He is the engineer. He holds his hand on the throttle, which allows the steam from the boiler to enter the cylinders in greater or less quantity, according to the speed he wishes to obtain. By one movement of the throttle, the steam is cut off from the cylinders and the engine stops; by another movement the steam is admitted and the locomotive moves, slowly or rapidly at will.

"The power of a locomotive is no doubt considerable; however, if it is able to draw with great speed a long train of cars, all heavily loaded, this is due, above all, to the preparation of the road on which it runs. Strong bars of iron, called rails, are fixed solidly on the road, all along its length, in two parallel lines, on which all the wheels of the train roll without ever running off. A light flange with which the wheels are furnished keeps the train from slipping off the rails.

"The iron road not having the inconveniences of other roads, that is to say the ruts, pebbles, and inequalities that impede the progress of carriages and cause the waste of much energy, the whole traction of the locomotive is utilized, and the results obtained are wonderful. A passenger engine draws at a rate of twelve leagues an hour a train weighing as much as 150,000 kilograms. A freight engine pulls at about seven leagues an hour a total weight of 650,000 kilograms. More than 1300 horses would be necessary to replace the first locomotive, and more than 2000 to replace the second, if they were employed to transport similar loads with the same velocity and to the same distances by the aid of cars running on rails. What an army of horses it would require with wagons running on ordinary roads having all the inequalities that cause such a great loss of energy!

"And now, my little friends, think of the thousands of locomotives running daily in all parts of the world, annihilating distances, as it were, and bringing the most distant nations together; think what a vast number of machines of all kinds, moved by steam, are ceaselessly working for man; think how the engine that makes a warship move, sometimes represents in itself the united strength of 42,000 horses; think of all these things, and see what inconceivable development of power man's genius has given to him with a few shovelfuls of coal burning under a pot of water!"

"Who first thought of the use of steam?" asked Jules. "I should like to remember his name."

"The use of steam as a mechanical power was proposed nearly two hundred years ago by one of the glories of France, the unfortunate Denis Papin, who, after giving the first suggestion of the steam-engine, source of incalculable riches, languished in a foreign land, poverty-stricken and forlorn. To realize his fruitful idea, which was to increase man's motive power a hundredfold, he could hardly find a paltry half-crown."

CHAPTER L

EMILE'S OBSERVATION

E MILE'S TURN came to tell what he had seen.

"When you made me a sign to be silent," said he, "it seemed to me as if the trees were walking. Those along the railroad were going very fast; farther away, the big poplars, ranged in long rows, were going with their heads waving as if saying good-by to us. Fields turned around, houses fled. But on looking closer I soon saw that we were moving and all the rest was motionless. How strange! You see something running that is really not moving at all."

"When we are comfortably seated in the railway car," his uncle replied, "without any effort on our part to go forward, how can we judge of our motion except by the position we occupy in relation to the objects that surround us? We are aware of the way we are going by the continual changing of the objects in sight, and not by any feeling of fatigue, since we do not move our legs. But the objects and people nearest to us and always before our eyes, our traveling companions and the furnishings of the car, remain for us in the same position. The left-hand neighbor is always at the left, the one in front is always in front. This apparent immobility of everything in the car makes us lose consciousness of our own movement; then we think ourselves immobile and fancy we see flying in an opposite direction exterior objects, which are always changing as we look at them. Let the train stop, and immediately trees and houses cease moving, because we no longer have a shifting point of view. A simple carriage drawn by horses, a boat borne along by the current, lend themselves to this same curious illusion. Every time we ourselves are gently moved along, we tend, more or less, to lose consciousness of this movement, and surrounding objects, in reality immobile, seem to us to move in a contrary direction."

"Without being able to explain it to myself well," returned Emile, "I see that it is so. We move and we think we see the other things moving. The faster we go, the faster the other things seem to go."

"You hardly suspect, my little friends, that Emile's naïve observation leads us straight to one of the truths that science has had the most trouble in getting accepted, not on account of its difficulty, but because of an illusion that has always deceived most people.

"If men passed their whole life on a railroad, without ever getting out of the car, stopping, or changing speed, they would firmly believe trees and houses to be in motion. Except by profound reflection, of which not everybody is capable, how could it be otherwise, since none would have seen the testimony of their eyes contradicted by experience? Of those that have been convinced, one sharper than the others rises and says this: 'You imagine that the mountains and houses move while you remain at rest. Well, it is just the opposite: we move and the mountains, houses, and trees stand still.' Do you think many would agree with him? Why! they would laugh at him, for each one sees, with his own eyes, mountains running, houses traveling. I tell you, my children, they would laugh at him."

"But, Uncle—" began Claire.

"There is no but. It has been done. They have done worse than laugh; they have become red with anger. You would have been the first to laugh, my girl."

"I should laugh at somebody asserting that the car moves and not the houses and mountains?"

"Yes, for an error that accompanies us all through life and that every one shares, is not so easily removed from the mind."

"It is impossible!"

"It is so possible that you yourself, at every turn, make the mountain move and the car that carries us stand still."

"I do not understand."

"You make the round earth, the car that bears us through celestial space, stand still; and you give motion to the sun, the giant star that makes our earth seem as nothing by comparison. At least, you say the sun rises, pursues its course, sets, and begins its course again the next day. The enormous star moves, the humble earth tranquilly watches its motion."

"The sun does certainly seem to us," said Jules, "to rise at one side of the sky and set at the other, to give us light by day. The moon does the same, and the stars too, to give us light at night."

"Listen then to this. I have read, I don't know where, of an eccentric person whose wrong-headedness could not reconcile him to simple methods. To attain the simplest result he would use means whose extravagance caused every one to laugh. One day, wishing to roast a lark, what do you think he took it into his head to do? I will give you ten, a hundred guesses. But, bah! you would never guess it. Just imagine! He constructed a complicated machine, with much wheelwork and many cords, pulleys, and counterpoises; and when it was started there was a variety of movement, back and forth, up and down. The noise of the springs and the grinding of the wheels biting on each other was enough to make one deaf. The house trembled with the fall of the counterpoises."

"But what was the machine for?" asked Claire. "Was it to turn the lark in front of the fire?"

"No, indeed; that would have been too simple. It was to turn the fire before the lark. The lighted firebrands, the hearth and chimney, dragged heavily by the enormous machine, all turned around the lark."

"Well, that beats all!" Jules ejaculated.

"You laugh, children, at this odd idea; and yet, like that eccentric man, you make the firebrands, hearth, the whole house turn around a little bird on the spit. The earth is the little bird; the house is the heavens, with their enormous, innumerable stars."

"The sun isn't very big—at most, as large as a grindstone," said Jules. "The stars are only sparks. But the earth is so large and heavy!"

"What did you just say? the sun as large as a grindstone? the stars only little sparks? Ah, if you only knew! Let us begin with the earth."

CHAPTER LI

A JOURNEY TO THE END OF THE
WORLD

" **A** SMALL BOY, of Jules's age and, like him, desirous to learn, one morning was making his preparations for a journey. Never had a navigator getting ready for a voyage over distant seas shown more zeal. Provisions, the first necessity in long expeditions, were not forgotten. Breakfast was doubled. There were in the basket six nuts, a bread-and-butter sandwich, and two apples! Where can one not go with all that? The family was not informed: they might have dissuaded the audacious traveler from his project by acquainting him with the perils of the expedition. For fear of softening before his mother's tears, he kept silent. Basket in hand, without saying good-by to any one, he takes his departure. Soon he is in the country. To left or right makes no difference to him; all roads lead whither he wishes to go."

"Where does he want to go?" asked Emile.

"To the end of the world. He takes the right-hand road, which is bordered by a hawthorn hedge where golden green beetles rustle and shine. But the beautiful insects do not stop him for a moment, nor yet the little red-bellied fish that play in the streamlet. The day is so short and the journey so long! He keeps on walking straight ahead, sometimes shortening the distance by cutting across fields. At the end of an hour the sandwich, chief item in the provisions, had been eaten, although the eating of it was regulated by the wise economy of a prudent traveler. Quarter of an hour later an apple and three nuts were gone. Appetite comes quickly to those who tire themselves. It comes so quickly that at a turn of the road, in the shade of a large willow, the second apple and the three remaining nuts are taken out of the basket. The provisions were exhausted, and (no less grave a matter) legs refused to go. Just imagine the situation. The journey had lasted two hours, and the end proposed was no nearer, not a bit. The little boy retraced his steps, persuaded that with better legs and more provisions he would succeed another time in his project."

"What was this project?" Jules asked.

"I told you: the audacious child wished to reach the end of the world. According to his ideas, the sky was a blue vault, which kept getting lower until it rested on the edge of the earth, so that, if ever he arrived there, he would have to walk bent over so as not to bump his head against the firmament. He started with the idea that he should soon be able to touch the sky with his hand; but the blue vault, retiring as he advanced, was always at the same distance. Fatigue and want of provisions made him renounce further continuance of his journey."

"If I had known that little boy," said Emile, "I would have dissuaded him from his expedition. It is impossible, however far one goes, to touch the sky with the hand, even with the help of the tallest ladder."

"If I remember aright, Emile has not always been of that opinion," said his uncle.

"That is true, Uncle. Like the little boy you have been telling about, I believed that the sky was a large blue cover resting on the earth. By good walking one ought to reach the edge of the cover and the end of the world. I thought, too, that the sun rose behind these mountains, and set behind those on the opposite side, where there was a deep well that the sun plunged into and remained hidden during the night. One day you took me to the mountains where the edges of the blue cover seem to rest. It was a long way off, I remember; you lent me your cane, which helped me in walking. I did not see any well for the sun to plunge into; everything looked just as it does here. The edge of the sky still seemed to rest on the earth, only much farther away. And you told me that by going to the end of what we saw, then farther and farther still, we should find the same appearance everywhere, without ever seeing the end of a vault that does not really exist."

"Nowhere, as all three of you know, does the sky rest on the earth; nowhere is there any danger of striking one's head against the firmament; everywhere the blue vault has the same appearance as here. You know, too, that in always going ahead you meet with plains, mountains, valleys, water-courses, seas; but nowhere are there any barriers marking the limits of the world.

"Imagine a large ball suspended in the air by a thread, and on this ball a gnat. If this gnat should take a notion to go all over the surface, is it not true that it could come and go over the ball, above, below, on the side, without ever encountering an obstacle, without ever seeing a barrier rise up to block its passage? Is it not equally true that if it always kept on in the same direction, the gnat would end by making the tour of the ball and would come back to its starting-point? So it is with us on the surface of the earth, though we are far more insignificant when compared with the globe that bears us than is the tiniest gnat in comparison with the biggest ball you can imagine. Without ever encountering a barrier, without ever touching the cupola of the sky, we come and go in a thousand different directions, we accomplish the most distant journeys, even make the tour of the earth and return to our starting-point. The earth, then, is round; it is an immense ball that swims without support in celestial space. As to the blue vault that arches above us, it is mere appearance caused by the blue color of the air enveloping the earth on all sides."

"The ball on which your imaginary gnat travels is suspended by a thread. By what chain is the enormous ball of the earth hung?" asked Jules.

"The earth is not suspended from the firmament by any celestial chain, nor does it rest upon any support, like a geographical globe on its pedestal. According to an Indian legend the terrestrial globe is borne upon four bronze columns."

"And what do the four columns rest on, in their turn?"

"They rest on four white elephants."

"And the white elephants?"

"They rest on four monstrous turtles."

"And the turtles?"

"Well, they swim in an ocean of milk."

"And the ocean of milk?"

"The legend says nothing about that, and it is right to be silent. It would have been better not to imagine all these various supports, resting one on another, to hold the earth up. Suppose a pedestal for the earth, then a second to uphold the first, then a third, fourth, thousandth, if you like; it is only postponing the question without answering it, since finally, after having erected all the supports imaginable, one must ask what will the last one rest on. Perhaps you are thinking of the vault of the heavens, which might well sustain the earth; but know that this vault has no reality, that it is nothing but an appearance caused by the air. Besides, thousands of travelers have gone over the earth in every direction, and nowhere have they seen either a suspending chain or a pedestal of any kind. Everywhere they see only what is to be seen here. The earth is isolated in space; it swims in a void without any support, just as do the moon and the sun."

"But, then, why doesn't it fall?" persisted Jules.

"To fall, my little friend, is to rush earthward as a stone does when raised in the hand and then left to itself. How can the large ball rush to the earth, when it is the whole earth? Is it possible for a thing to rush toward itself?"

"No."

"Well, then! Besides, imagine this. All is the same around the terrestrial globe; properly speaking, there is no up or down, no right or left. We call up the direction toward adjacent space, or toward the sky; but remember that there is sky also on the other side of the earth, that there it is just the same as we see it here, and that this is true for all parts of the earth's surface. If it seemed to you quite simple that the earth does not rush toward the sky which is above us, why should you expect it to rush toward the opposite sky? To fall toward the opposite sky would be to rise, as the lark rises here, when with one stroke of the wing it takes its flight and soars above us."

CHAPTER LII

THE EARTH

"THE EARTH is round, as proved by the following facts. When, in order to reach the town he is journeying toward, a traveler crosses a level plain where nothing intercepts his view, from a certain distance the highest points of the town, the summits of towers and steeples, are seen first. From a lesser distance the spires of the steeples become entirely visible, then the roofs of buildings themselves; so that the view embraces a

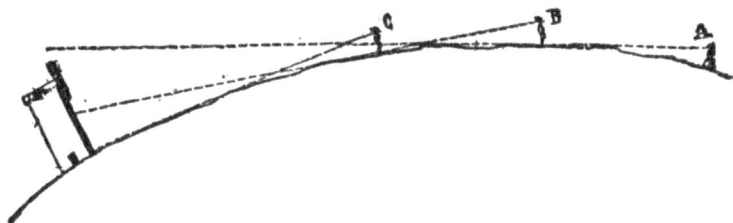

great number of objects, beginning with the highest and ending with the lowest, as the distance diminishes. The curvature of the ground is the cause of it."

Uncle Paul took a pencil and traced on paper the picture that you see here; then he continued:

"To an observer at A the tower is quite invisible because the curvature of the ground hides the view. To the observer at B the upper half of the tower is visible, but the lower half is still hidden. Finally, when the observer is at C he can see the whole tower. It would not be thus if the earth were flat. From any distance the whole of a tower would be visible. Afar off, no doubt, it would be seen with less clearness than near to, on account of the distance; but it could be seen more or less well from top to bottom."

Here is another drawing of Uncle Paul's, representing two spectators, A and B, who, placed at very different distances, nevertheless see the tower from top to bottom on a flat surface. He resumed his talk.

"On dry land it is rare to find a surface that in extent and regularity is adapted to the observation I have just told you about. Nearly always hills, ridges, or screens of verdure intercept the

view and prevent one's seeing the gradual appearance, from summit to base, of the tower or steeple that one is approaching. On the sea no obstacle bars the view unless it be the convexity of the waters, which follow the general curvature of the earth. It is, accordingly, there especially that it is easy to study the phenomena produced by the rounded form of the earth.

"When a ship coming from the open sea approaches the coast, the first points of the shore visible to those on board are the highest points, like the crests of mountains. Later the tops of high towers come into sight; still later the edge of the shore itself. In the same way an observer who witnesses from the shore the arrival of a vessel begins by perceiving the tops of the masts, then the topsails, then the sails next below, and finally the hull of the vessel. If the vessel were departing from the shore, the observer would see it gradually disappear and apparently plunge under the water, all in inverse order; that is to say, the hull would be first hidden from view, then the low sails, then the high ones, and finally the top of the mainmast,

which would be the last to disappear. Four strokes of the pencil will make you understand it."

"How large is the earth?" was the next question from Jules.

"The earth is forty million meters in circumference or 10,000 leagues, for a league measures four kilometers. To encircle a round table, you take hold of hands, three, four, or five of us. To encircle in the same manner the vast bosom of the earth, it would take a chain of people about equal to the whole population of France. A traveler able to walk day after day at the rate of ten leagues a day, which no one could do, would take three years to girdle the globe, supposing it to be all land and no sea. But, where are the hamstrings that could resist three years of such continual fatigue, when a walk of ten leagues generally exhausts our strength and makes it impossible for us to begin again the next morning?"

"The longest walk I ever took was to the pine woods, where we went to look for the nest of the processionary caterpillars, the day of the thunderstorm. How many leagues did we go?"

"About four, two to go and two to come back."

"Only four leagues! All the same I was played out. At the end I could hardly put one foot before the other. It would take me, then, from seven to eight years to go round the world, walking every day as far as my strength would let me."

"Your calculation is right."

"The earth then is a very large ball?"

"Yes, my friend, very large. Another example will help you to understand it. Let us represent the terrestrial globe by a ball of greater diameter than a man's height—by a ball two meters in diameter; then, in correct proportion, represent in relief on its surface some of the principal mountains. The highest mountain in the world is Gaurisankar, a part of the Himalaya chain, in central Asia. Its peaks rise to a height of 8840 meters. Rarely are the clouds high enough to crown its crest, and its base covers the extent of an empire. Alas! what does man become, materially, in face of such a prodigious colossus! Well, let us raise the giant on our large ball representing the earth; do you know what will be needed to represent it? A tiny little grain of sand which would be lost between your fingers, a grain of sand that would stand out in relief only a millimeter and a third! The gigantic mountain that overwhelmed us with its immensity is nothing when compared with the earth. The highest mountain in Europe, Mont Blanc, whose height is 4810 meters, would be represented by a grain of sand half as large as the other."

"When you told us of the roundness of the earth," put in Claire, "I thought of the enormous mountains and deep valleys, and asked myself how, with all these great irregularities, the earth could nevertheless be round. I see now that these irregularities are a mere nothing in comparison with the immensity of the terrestrial ball."

"An orange is round in spite of the wrinkles in its skin. It is the same with the earth: it is round in spite of the irregularities of its surface; it is an enormous ball sprinkled with grains of dust and sand proportioned to its size, and these are mountains."

"What a big ball!" exclaimed Emile.

"To measure the circumference of the earth is not an easy thing, you may be sure; and yet they have done more than that: they have weighed the immense ball as if it were possible to put it in a scale-pan with kilograms for counterweights. Science, my dear children, has resources demonstrating in all its grandeur the power of the human mind. The immense ball has been weighed. How it was done cannot be explained to you to-day. No scales were used, but it was accomplished by the power of thought with which God has endowed us, to solve, to His glory, the sublime enigma of the universe; by the force of reason, for which the burden of the earth is not too heavy. This burden is expressed by the figure 6 followed by twenty-one zeros, or by 6 sextillions of kilograms."

"That number means nothing to me; it is too large," Jules declared.

"That is the trouble with all large numbers. Let us get around the difficulty. Suppose the earth placed on a car and drawn on a surface like that of our roads. For such a load, what should the team be? Let us put in front a million horses; and in front of that row a second million; then a third row, still of a million; a hundredth, finally a thousandth. We shall thus have a team of a thousand millions of horses, more than could be fed in all the pastures of the world. And now start; apply the whip. Nothing would move, my children; the power would be insufficient. To start the colossal mass, it would need the united efforts of a hundred millions of such teams!"

"I don't grasp it any better," said Jules.

"Nor I, it is so enormous," assented his uncle.

"Yes, enormous, Uncle."

"So that the mind gets lost in it," said Claire.

"That is what I wanted to make you acknowledge," concluded Uncle Paul.

CHAPTER LIII

THE ATMOSPHERE

"I F YOU pass your hand quickly before your face, you feel a breath blow on your cheeks. This breath is air. In repose it makes no impression on us; put in motion by the hand, it reveals its presence by a light shock that produces an impression of freshness. But the shock from the air is not always, like this, a simple caress. It can become very brutal. A violent wind, which sometimes uproots trees and overthrows buildings, is still air in motion, air that flows from one country to another like a stream of water. Air is invisible, because it is transparent and almost colorless. But if it forms a very thick layer through which one can look, its feeble coloring becomes perceptible. Seen in small quantities, water appears equally colorless; seen in a deep layer, as in the sea, in a lake, or in a river, it is blue or green. It is the same with air: in thin strata it seems deprived of color; in a layer several leagues in thickness, it is blue. A distant landscape appears to us bluish, because the thick bed of intervening air imparts to it its own color.

"Now air forms all around the earth an envelope fifteen leagues thick. It is the aërial sea or atmosphere, in which the clouds swim. Its soft blue tint causes the sky's color. It is in fact the atmosphere that produces the appearance of a celestial vault.

"Do you know, my children, what is the use of this aërial sea at the bottom of which we live as fish live in water?"

"Not very well," Jules replied.

"Without this ocean of air life would be impossible, plant life as well as animal. Listen. Chief of those imperious needs to which we are subjected are those of eating, drinking, and sleeping. As long as hunger is only its diminutive, appetite, that

savory seasoning of the grossest viands; as long as thirst is only
that nascent dryness of the mouth that gives so great a charm
to a glass of cold water; as long as sleepiness is nothing more
than that gentle lassitude that makes us desire the night's rest,
so long is it the attraction of pleasure rather than the rude prick
of pain that urges the satisfaction of these primordial needs. But
if their satisfaction is too long delayed, they impose themselves
as inexorable masters and command by torture. Who can think
without terror of the agonies of hunger and thirst! Hunger! Ah!
you do not know what it is, my children, and God preserve you
from ever knowing it! Hunger! If you could have any idea of its
tortures, your heart would be oppressed at the thought of the
unhappy ones who experience it. Ah! my dear children, always
help those that are hungry; help them, and give, give; you will
never do a nobler deed in this world. Giving to the poor is lending to the Lord."

Claire had put her hand before her eyes to hide a tear of
emotion. She had observed a flash on her uncle's face that spoke
from the depth of his heart. After a moment's pause Uncle Paul
continued:

"There is, however, a need before which hunger and thirst,
however violent they may be, are mute; a need always springing
up afresh and never satisfied, which continually makes itself felt,
awake or asleep, night or day, every hour, every moment. It is
the need of air. Air is so necessary to life that it has not been
given us to regulate its use, as we do with eating and drinking,
so as to guard us from the fatal consequences that the slightest
forgetfulness would cause. It is, as it were, without consciousness or volition on our part that the air enters our body to
perform its wonderful part. We live on air more than anything
else; ordinary nourishment comes second. The need of food is
only felt at rather long intervals; the need of air is felt without
ceasing, always imperious, always inexorable."

"And yet, Uncle," said Jules, "I have never thought of feeding myself with air. It is the first time I ever heard that air is so
necessary for us."

"You have not given it a thought, because all that is done for
you; but try a moment to prevent air entering into your body:

close the ways to it, the nose and mouth, and you will see!"

Jules did as his uncle told him, shut his mouth and pinched his nose with his fingers. At the end of a moment, his face red and puffed up, the little boy was obliged to put an end to his experiment.

"It is impossible to keep it up, Uncle; it suffocates a person and makes him feel as if he should certainly die if it kept on a little longer."

"Well, I hope you are convinced of the necessity of air in order to live. All animals, from the tiniest mite, hardly visible, to the giants of creation, are in the same condition as you: on air, first of all, their life depends. Even those that live in the water, fish and others, are no exception to this rule. They can live only in water into which air infiltrates and dissolves. When you are older you shall see a striking experiment which proves how indispensable to life is the presence of air. You put a bird under a glass dome, shut tight everywhere; then with a kind of pump the air is drawn out. As it is withdrawn from the inside of the glass cage, the bird staggers, struggles a moment in an anguish horrible to see, and falls dead."

"It must take a lot of air," was Emile's comment, "to supply the needs of all the people and animals in the world. There are so many!"

"Yes, indeed; a great quantity is needed. One man needs nearly 6000 liters of air an hour. But the atmosphere is so vast that there is plenty of air for all. I will try to make you understand it.

"Air is one of the most subtle of substances; a liter of it weighs only one gram and three decigrams. That is very little: the same volume of water weighs 1000 grams; that is to say, 769 times as much. However, such is the enormous extent of the atmosphere that the weight of all the air composing it outstrips your utmost powers of imagination. If it were possible to put all the air of the atmosphere into one of the pans of an immense pair of scales, what weight do you think it would be necessary to put into the other pan to make it equal the air? Don't be afraid of exaggerating; you can pile up thousands on thousands of kilograms; if air is very light, the aërial sea is very vast."

"Let us put on a few millions of kilograms," suggested Claire.

"That is a mere trifle," her uncle replied.

"Let us multiply it by ten, by a hundred."

"It is not enough, the pan would not be raised. But let me tell you the answer, for in this calculation numerical terms would fail you. For the great weight I am supposing, the heaviest counterweights would be insignificant. New ones must be invented. Imagine, then, a copper cube, a kilometer in each dimension; this metallic die, measuring a quarter of a league on its edge, shall be our unit of weight. It represents nine thousand millions of kilograms. Well, to balance the weight of the atmosphere, it would be necessary to put into the other pan 585,000 of these cubes!"

"Is it possible!" Claire exclaimed.

"I told you so! Imagination vainly seeks to picture the stupendous mass of the layer of air wound like a scarf by the Creator around the earth. Now do you know what relation it bears to the terrestrial globe—this ocean of air having a weight represented by half a million of copper cubes a quarter of a league each way? Scarcely what the imperceptible velvety down of a peach is to the peach itself. What, then, are we, materially, we poor beings of a day, who move about at the bottom of this atmospheric sea! But how great we are through thought, which makes game of weighing the atmosphere and the earth itself! In vain does the material universe overwhelm us with its immensity; the mind is superior to it, because it alone knows itself, and it alone, by a sublime privilege, has knowledge of its divine author."

CHAPTER LIV

THE SUN

EARLY IN the morning Uncle Paul and his nephews climbed the neighboring hill to see the sunrise. It was still quite dark. The only persons they met in passing through the village were the milkmaid. on her way to town with her butter and milk, and the blacksmith hammering away at the red-hot iron on his anvil, while the glow from the forge illumined the darkness of the road.

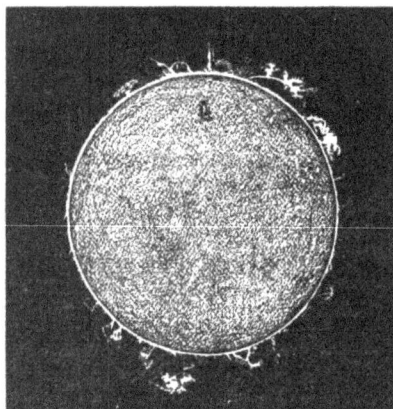

Sheltered by a clump of juniper-trees, Paul and the three children await the grand spectacle they have come to the top of the hill to see. In the east the sky is getting lighter, the stars turn pale and go out one by one. Flakes of rosy cloud swim in a brilliant streak of light whence gradually there rises a soft illumination. It reaches the zenith, and the blue of day reappears with all its delicate transparency. This cool morning light, this half-daylight that precedes the rising of the sun, is the aurora or morning twilight. In the meantime a lark, the joy of the fields, takes wing to the highest clouds, like a rocket, and is the first to salute the awakening day. It mounts and mounts, always singing, as if to get in front of the sun; and with its enthusiastic songs it celebrates in the high heavens the glory of the day-bringer. Listen: there is a breath of wind in the foliage, which stirs and rustles; the little birds are waking

THE SUN

up and chirping; the ox, already led to work in the fields, stops as if thinking, raises its large eyes full of gentleness, and lows; everything becomes animated, and, in its own language, renders thanks to the Master of all things, who with His powerful hand brings us back the sun.

And here it is: a bright thread of light bursts forth, and the tops of the mountains are suddenly illumined. It is the edge of the sun beginning to rise. The earth trembles before the radiant apparition. The shining disc keeps rising: there it is almost whole, now completely so, like a grindstone of red-hot iron. The mist of the morning moderates its glare and allows one to look it in the face; but soon no one could endure its dazzling splendor. In the meantime its rays inundate the plain; a gentle heat succeeds the keen freshness of the night; the mists rise from the depths of the valleys and are dissipated; the dew, gathered on the leaves, becomes warm and evaporates; on all sides there is a resumption of life, of the animation suspended during the night. And all day, pursuing its course from east to west, the sun moves on, flooding the earth with light and heat, ripening the yellow harvest, giving perfume to the flowers, taste to fruit, life to every creature.

Then Uncle Paul, in the shade of the juniper-trees, began his talk.

"What is the sun? Is it large, is it very far away? That, my children, is what I should now like to teach you.

"To measure the distance from one point to another, you know of only one means: that of laying off, as many times as it will go, the unit of length, the meter, from one end to the other of the distance to be measured. But science has methods adapted to the measuring of distances that one cannot travel in person; it tells us what must be done to find the height of a tower or mountain, without going to the top, without even approaching the base. They are methods of the same kind as are employed to calculate the distance that separates us from the sun. The result of the astronomer's calculations is that we are distant from the sun 38 millions of leagues of 4000 meters each. This distance is equivalent to 3800 times the circumference of the earth. I told you that, to make the tour of the terrestrial globe, a man, a good walker, capable of walking ten leagues a day, would take about three years. He would need, then, nearly twelve thousand years to go from the earth to the sun, supposing that the journey were possible. The longest human life is incomparably too short for a journey of this length ever to be accomplished by one person; and a hundred generations of a hundred years each, succeeding one another on the journey and uniting their efforts, would not even be enough."

"And a locomotive," asked Jules, "how long would it take to get over that distance?"

"Do you remember how fast it goes?"

"I saw it myself the day we took the trip with you. If one looks out, the road seems to fly back so fast it frightens you and makes you dizzy."

"The locomotive that drew us went at the rate of about ten leagues an hour. Let us suppose a locomotive that never stops and that goes still faster, or fifteen leagues an hour. Rushing at that speed, the engine would go from one end of France to the other in less than a day; and yet, to cover the distance from the earth to the sun, it would take more than three centuries. For such a journey, the fastest engine ever made by the hand of man is hardly more than a sluggish snail ambitious to make the tour of the world."

"And I who thought, not long ago," said Emile, "that by climbing to the roof and with the aid of a long reed I could touch the sun!"

"To one who trusts to appearances the sun is only a dazzling disc, at the most as large as a grindstone."

"That is what I said yesterday," observed Jules. "But, as it is so far away, it might well be as large as a millstone."

"In the first place, the sun is not flat like a grindstone; it has, like the earth, the shape of a ball. Furthermore, it is much larger than a grindstone, or even than a millstone.

"Objects seem to us small in proportion to their distance from us, until finally they become invisible. A high mountain seen from afar seems only a moderate-sized hill; the cross that surmounts a steeple, seen from below, looks very small despite its very large dimensions. It is the same with the sun: it looks so small only because it is very far off; and as the distance is prodigious, its size must be excessive; if not, instead of looking to us like a dazzling grindstone, it would cease to be visible to us.

"You found the terrestrial globe enormous; and, despite my comparisons, your imagination, I am sure, has not been able to picture things properly. How will it be with the sun, which is one million four hundred thousand times as large as the earth! If we suppose the sun hollow like a spherical box, to fill it would take one million four hundred thousand balls the size of the earth.

"Let us try another comparison. To fill the measure of capacity called the liter, it takes about 10,000 grains of wheat. It would take, then, 100,000 to fill 10 liters or one decaliter, and 1,400,000 to fill 14 decaliters. Well, suppose in one pile 14 decaliters of wheat, and beside it one solitary grain of wheat. For the respective sizes, this isolated grain represents the earth; the pile of 14 decaliters represents the sun."

"How wrong we were!" Claire exclaimed. "This little shining disc, to which, for fear of exaggeration, we should have hesitated to assign the dimensions of a millwheel, is a globe so big that in comparison with its gigantic size the earth is as nothing."

"Oh, God in heaven!" cried Jules.

"Yes, my friend, you may well say, 'God in heaven,' for the mind is bewildered at the thought of this inconceivable mass. Say: God in heaven! how great You are, You who out of nothing have created the sun and the earth, and hold them both in the shadow of Your hand!

"I have not finished, my dear children. One day, in speaking to you of lightning and thunder, I told you that light moves with excessive rapidity. In fact, to come to us from the sun, to cover the distance that a locomotive at its highest speed would take three hundred years to cover, a ray of light needs only the half of a quarter of an hour, or about eight minutes. Now listen to this. Astronomy teaches us that each star, small as it may appear from here, is itself a sun comparable in size to ours; it tells us that these suns, of which we with the naked eye can perceive only a very small part, are so numerous that it is impossible to count them; it tells us that their distance is so great that, to come to us from the nearest star, light, which travels so fast, as I have just told you, takes nearly four years; to reach us from others that are by no means the most distant it takes whole centuries. After that, if you can, estimate the distance that separates us from those far-off suns; think also of their number and size. But no, do not try: the intellect is overwhelmed by these immensities in which is revealed all the majesty of God's handiwork. Do not try, it would be in vain; but let arise from your heart the burst of admiration that you cannot suppress, and bless God, whose infinite power has scattered suns through the boundless regions of celestial space."

CHAPTER LV

DAY AND NIGHT

"IT SEEMS to me," said Claire, "we have lost sight of the hearth that turns with its lighted firebrands around the lark."

"On the contrary, we are closer to it than ever. If the sun, which is thirty-eight millions of leagues from us, were to go around the earth every day, do you know how far it would have to go in a minute? More than 100,000 leagues. But this incomprehensible speed is nothing. The stars, as I have just told you, are so many suns, comparable to ours in volume and brilliancy; only they are much farther away, and that is what makes them appear so small. The nearest is about thirty thousand times as distant as the sun. Accordingly, in order to go around the earth in twenty-four hours, as it appears to do, it would have to move at the rate of thirty thousand times 100,000 leagues a minute. And how would it be with other stars a hundred times, a thousand times, a million times farther away—stars which, despite their distance, would all have to accomplish their journey around the earth always in exactly twenty-four hours? And remember, furthermore, the prodigious size of the sun. You want it, the giant, the colossus, beside which the earth is only a lump of clay, to circle at an impossible speed in infinite space, in order to give light and heat to our planet; you want thousands and thousands of other suns, quite as large and immensely farther off—in a word, the stars—to accomplish also, with velocities increasing according to the distance, a daily journey around this humble terrestrial globe! No! no! such an arrangement is contrary to reason; to allow it is to want to make the firebrands, the hearth, the whole house, turn around a little bird on a spit."

"Then it is the earth that turns, and we turn with it," Claire again interposed. "In consequence of this movement the sun and stars seem to us to move in the opposite direction, like trees and houses when we are on the train. Since the sun seems to go around the earth from east to west in twenty-four hours, it is a proof that the earth turns on its axis from west to east in twenty-four hours."

"The earth turns in front of the sun in a manner to present its different parts successively to the rays of that body; it pirouettes on its axis like a top. Moreover, while it thus rotates in twenty-four hours, it revolves around the sun in the interval of a year. In playing with a top you find a good example of two analogous movements executed together. When the top turns on its point, not moving from the same place—in short, when it sleeps—it has only the movement of rotation. But in throwing it in a certain way, you know better than I that it circles on the ground while turning on its point. In that instance, it represents in a small way the double movement of the earth. Its rotation on its point represents the whirling motion of the earth on its axis; its course on the ground represents the earth's revolution around the sun.

"You can familiarize yourself in another way with the double movement of the terrestrial globe, as follows: place in the middle of a room a round table, and on that table a lighted candle to represent the sun. Then circle around the table, pirouetting on your toes. Each of your pirouettes corresponds to a turn of the earth on its axis, and your course around the table corresponds to its journey around the sun. Notice that in turning on your toes you present in succession to the rays of the candle the front, one side, the back, and the other side of your head, which in our experiment may represent the terrestrial globe; so that each one of its parts is in turn in the light or in the shade. The earth does the same: in turning it presents one after the other its different regions to the rays of the sun. It is day for the region that sees the sun, night for the opposite region. That is the very simple cause of day and night. In twenty-four hours the earth makes one rotation on its axis. Of these twenty-four hours the duration of the day and night is composed."

"I understand very well the cause of the alternation of day

and night," said Jules. "It is day for the half of the earth that
sees the sun, night for the opposite half. But as the globe turns,
each country comes in succession to face the sun while others
pass into the unlighted half. The lark that turns on the hearth
presents, in the same way, each of its sides in turn to the heat
of the flame."

"One might almost say," remarked Emile, "it is day for the
half of the lark next to the fire, and night for the other half."

"One difficulty still perplexes me," Jules continued. "If the
earth turns around once in every twenty-four hours, in half
of that time we ought to make a half-turn with the globe that
carries us, and find ourselves upside-down. At this moment we
have our heads up, feet down; twelve hours later it will be just
the opposite: our heads will be down and our feet up. We are
upright, we shall be upside-down. In that inconvenient position why don't we feel uncomfortable? Why are we not thrown
down? So as not to fall, it seems to me, we ought to be obliged
to cling to the ground in desperation."

"Your observation is right," returned Uncle Paul, "but only
in a certain degree. Yes, it is true that twelve hours from now
we shall be in an inverse position; our heads will be toward that
point in space to which our feet are now turned. But despite this
inversion there will be no danger of our falling, nor even the
slightest inconvenience of any kind; for our heads will always
be up, that is to say toward the sky, since the sky surrounds the
terrestrial globe everywhere; our feet will always be down, that
is to say resting on the ground. Understand thoroughly, once
for all, that to fall is to rush toward the ground, and not into
surrounding space. So that notwithstanding all the evolutions
of our globe, as we are always on the earth, feet on the ground,
head toward the sky, we are always in an upright position,
without any unpleasant feeling, without any danger of falling."

"Does the terrestrial globe turn very fast?" Emile inquired.

"It turns on its axis once in twenty-four hours. Therefore
any point in its middle region, the region that makes the longest journey, travels in the same time forty millions of meters,
that is to say a journey equal to the circuit of the earth, or 462
meters a second. That is about the speed of a cannon-ball as it

leaves the cannon's mouth, or about thirty times the speed of the fastest locomotive. Mountains, plains, seas, apparently fixed in their places for time and for eternity, are perpetually chasing one another in a circle, with the formidable speed of more than one-tenth of a league a second."

"And yet everything seems to us to be stationary."

"Without the jolting of the car should we not think we were standing still when the train carries us with such frightful speed? Well, the rapid movement of the earth is at the same time so gentle that it is impossible to be aware of it except by the apparent motion of the stars."

"By rising to a certain height in a balloon," said Jules, "we ought to see the earth turning under us. Seas and their islands, continents with their empires, forests, and mountains, ought in succession to come under the eyes of the observer, who in twenty-four hours sees the turning of the whole earth. What a magnificent spectacle that must be! What a journey, so wonderful and with so little fatigue! When the rotation brings back one's own country, one descends and it is accomplished. In twenty-four hours, without changing place, one has seen the whole world."

"Yes, I agree with you, it would be an admirable way to see countries. To this spot where we are other peoples will come, brought by the rotation; seas, distant regions, snowy mountains will take our place; and to-morrow at the same hour we shall be here again. Where we are talking now, in the shade of the juniper-trees, first will pass the sea, the somber Atlantic, which will replace our conversation by the grand voice of its waves. In less than an hour the ocean will be here. Some large war-vessel, with its triple row of guns, will float perhaps, all sails set, over the spot we are occupying. The sea has passed. Now we have North America, the great Canadian lakes, and the interminable prairies where the red-skinned Indians hunt buffaloes. The sea begins again, much larger than the Atlantic; it takes nearly seven hours to pass. What line of islands is this where fishermen wrapped in furs are drying herrings? They are the Koorile Isles, south of Kamchatka. They pass quickly; we scarcely have time to give them a glance. Now it is the turn of the yellow-faces—the Mongolians and Chinese, with slanting eyes. Oh! what curious things we could see here! But the ball is always turning, and China is already in the distance. The sandy plateaus of Central Asia and mountains higher than the clouds come next. Here are the pastures of the Tartars, with neighing herds of mares; here are the grassy plains of the Caspian with the flat-nosed Cossacks; then southern Russia, Austria, Germany, Switzerland, and finally France. Let us descend quickly, get on to our feet; the earth has finished its rotation.

"Do not for an instant, my little friends, think that this giddy spectacle of the earth passing with the rapidity of a cannon-ball would be visible to any but spiritual eyes. By rising into the upper air in a balloon, as Jules said, it does at first seem as if we ought to see the earth turning and lands and seas passing under our feet. Nothing of the kind takes place, for the atmosphere turns with the terrestrial globe and drags the balloon in the general rotation, instead of leaving it at rest, as would be necessary if the observer were to have successively under his eyes the different regions of the earth."

CHAPTER LVI

THE YEAR AND ITS SEASONS

"Y OU TOLD us," said Claire, "that at the same time the earth turns on its axis it travels round the sun."

"Yes. It takes three hundred and sixty-five days for that journey; it makes three hundred and sixty-five pirouettes on its axis in accomplishing a journey round the sun. The time spent in this journey makes just a year."

"The earth takes one day of twenty-four hours to turn on its axis; one year to turn round the sun," said Jules.

"That is it. Imagine yourself turning around a circular table the center of which is occupied by a lamp representing the sun, while you represent the earth. Each of your walks around the table is one year. To represent things exactly, you must turn on your heels three hundred and sixty-five times while you circle the table once."

"It is as if the earth waltzed around the sun," Emile suggested.

"The comparison is not so well chosen as it might be, but it is exact. It shows that in spite of the giddiness of his age Emile has understood perfectly. A year is divided into twelve months which are: January, February, March, April, May, June, July, August, September, October, November, December. The unequal length of the months is sometimes confusing. Some have 31 days, others 30; February has 28 or 29, according to the year."

"For my part," said Claire, "I should find it hard to tell whether May, September, and other months have 30 or 31 days. How can one remember which months have 31 days and which 30?"

"A natural calendar, engraved on our hands, teaches us in a very simple way. Close the fist of the left hand. At the knuckles the four fingers, other than the thumb, from each a bump, separated by a hollow from the next bump. Place the index finger of the right hand in turn on these bumps and hollows, beginning with the little finger, and at the same time name the months of the year in order: January, February, March, etc. When the series of the four fingers is exhausted, return to the starting-point and continue naming the twelve months on the bumps and hollows. Well, all the months corresponding to the bumps have 31 days; all those corresponding to the hollows, 30. You must except February, answering to the first hollow. That has 28 or 29 days, according to the year."

"Let me try," proposed Claire. "We'll see how many days May has: January, bump; February, hollow; March, bump; April, hollow; May, bump. May has 31 days."

"It is as easy as that," said her uncle.

"My turn now," interposed Jules. "Let us try September: January, bump; February, hollow; March, bump; April, hollow; May, bump; June, hollow; July, bump. And now? I am at the end of my hand."

"Now begin again and go on naming the months," Uncle Paul instructed him.

"You go on at the same point where you began?"

"Yes."

"All right. August, bump. There are two bumps in succession. There are then two months together, July and August, that have 31 days?"

"Yes."

"I will begin again. August, bump; September, hollow. September has 30 days."

"Why has February sometimes 28 and sometimes 29 days?" asked Claire.

"I must tell you that the earth does not take exactly 365 days to turn around the sun. It takes nearly six hours more. To make up these six hours that were disregarded at first in order to have a round number of days in the year, they are reckoned in every four years, and the additional day they make all together is added to February, which then becomes 29 days long instead of 28."

"So, for three years running, February has 28 days, and the fourth year it has 29."

"Exactly. Remember, too, that the years when February has 29 days are called leap years."

"And the seasons?" queried Jules.

"For reasons that would be a little too difficult for you to understand yet, the annual journey of the earth around the sun causes the seasons and the unequal length of days and nights.

"There are four seasons, of three months each: spring, summer, autumn, and winter. Spring is from about March 20th to June 21st; summer from June 21st to September 22nd; autumn from September 22nd to December 21st; winter from December 21st to March 20th.

"On March 20th and September 22nd the sun is visible 12 hours and invisible 12 hours, from one end of the earth to the other. The 21st of June is for us the time of the longest days and shortest nights; the sun is visible sixteen hours and invisible eight hours. Farther north the length of the day increases and that of the night diminishes. There are countries where the sun, an earlier riser than here, rises at two o'clock in the morning and sets at ten o'clock at night; still others where the time of its rising and that of its setting are so close together that the sun has hardly sunk below the apparent edge of the sky before it appears again. Finally, at the very pole of the earth, that is to say at the point that remains stationary, like the end of the axle of a wheel, while all the rest turns, one could witness the wonderful spectacle of a sun that does not set, that turns around the spectator for six whole months, equally visible at midnight and midday. In those countries there is no longer any night.

"On the 21st of December we have a state of affairs just the reverse of that observed in June. With us the sun rises at 8 o'clock in the morning; at four in the afternoon it has already set. That is eight hours of day for sixteen of night. Farther north there are now nights of 18, 20, 22 hours, and corresponding days of six, four, and two hours. In the neighborhood of the pole, the sun does not even show itself, and there is no longer any daylight; for six months there is the same darkness in the middle of the day as at midnight."

"And do people live in that country of the pole, where the year is composed of a day lasting six months and a night of six months?" asked Jules.

"No, up to this time[1] man has not been able to reach the pole on account of the horrible cold there; but there are countries more or less near the pole which are inhabited. When winter comes, wine, beer, and other beverages turn into blocks of ice in their casks; a glass of water thrown into the air falls back in flakes of snow; the moisture of the breath becomes needles of rime at the opening of the nostrils; the sea itself freezes to a great depth and thus increases the apparent extent of the dry land, which it resembles, having, like it, its fields of snow and mountains of ice. For whole months the sun does not show itself, and there is no difference between day and night, or rather it is one long night, the same at midday as at midnight. However, when the weather is fine darkness is not complete; the light of the moon and stars, augmented by the whiteness of the snow, produces a kind of semi-daylight sufficient for seeing. By this wan light, in sledges drawn in disorderly fashion by teams of dogs, the people of these dark regions hunt what scanty game there is. Fishing furnishes them more abundant food. Fish, dried, stored, half decayed, and rancid whale's blubber are their habitual food. For fuel for their hearths their dependence is, again, on their fishing, which supplies them with fish-bones and slices of blubber. Here, in short, wood is unknown; no tree, however hardy, can resist the rigors of winter. Willows, birches, dwarfed to insignificant underbrush, venture as far as the southern extremities of Lapland, where the cultivation of barley, the hardiest of cultivated plants ceases. Beyond this point all woody vegetation ceases; and during the summer there are found only occasional tufts of grass and

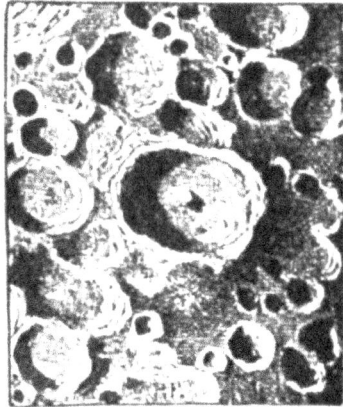

A PART OF THE MOON'S SURFACE

moss, hastily ripening their seeds in the sheltered hollows of

---

1   This was written before Peary's and Amundsen's achievements in polar exploration.—Translator.

the rocks. Further on the summer is too short for the snow and ice to melt completely; the ground is never bare, and all vegetation is impossible."

"Oh, the doleful countries!" cried Emile. "One more question, Uncle. In traveling around the sun does the earth go fast?"

"It takes a year for the entire tour; but as it circles at an enormous distance from the sun, a distance of 38 millions of leagues, it must travel this wide circle with a speed beyond your power to conceive. This speed is 27,000 leagues an hour. In the same time the fastest locomotive goes about 15 leagues. Compare and judge."

"What!" exclaimed Jules, "the immense ball of which we have never been able to comprehend the frightful weight travels in the sky with such rapidity?"

"Yes, my friend; with a speed of twenty-seven thousand leagues an hour the terrestrial ball goes rolling through space, without axle, without support, always on the ideal line that has been given it for its race-track. Who caused it to move so rapidly that the very thought of it makes you feel giddy? Let us bow the head, my children; it is the power of God."

CHAPTER LVII

BELLADONNA BERRIES

BAD NEWS was circulating from house to house in the village. Here is what they were saying:

That day they had put little Louis into his first trousers. They had pockets and shiny buttons. In his new costume Louis was a little awkward, but much pleased. He admired the buttons that shone in the sun; he kept turning his pockets inside out to see if there was room enough for all his playthings. What made him the happiest was a tin watch, always marking the same hour. His brother, Joseph, two years older, was also much pleased. Now that Louis was dressed like him, nothing prevented his taking him to the woods, where there were birds' nests and strawberries. They owned in common a lamb whiter than snow, with a pretty little bell at its neck. The two brothers were to take it to the meadow. Some lunch was packed in a basket. They kissed their mother, who advised them not to go far. "Take care of your brother," said she to Joseph; "hold him by the hand and come back soon." They started. Joseph carried the basket, Louis led the lamb. From the door their mother watched them going off, herself happy in their joy. Every now and then the children turned to smile at her; then they disappeared at the turn of the path.

They reach the meadow. The lamb frolics on the grass; Joseph and Louis run after butterflies in the midst of a clump of tall trees.

"Oh, the beautiful cherries!" exclaimed Louis, suddenly; "see how big and black they are! Cherries, cherries! We are going to have a feast. Let us pick some to eat."

There were, in fact, some large berries of a dark violet hue on low plants.

"How small these cherry-trees are!" answered Joseph. "I have never seen any like them. We shan't have to climb the tree for them, and you won't tear your new trousers."

Louis picked one of the berries and put it into his mouth. It was insipid and sweetish.

"These cherries are not ripe," says little Louis, spitting it out.

"Take this one," answers Joseph, giving him one that felt very soft. "It is ripe."

Louis tastes it and spits it out.

"No, they are not at all good," repeats the little boy.

"Not good, not good?" says Joseph; "you will see." He eats one, then another, then another still, then a fourth, then a fifth. At the sixth he is obliged to stop. Decidedly they were not good.

"It is true, they are not very ripe. But let's pick some, all the same. We'll let them ripen in the basket."

They gathered a handful or two of these black berries, then began running after butterflies. The cherries were forgotten.

An hour later, Simon, who was returning from the mill with his donkey, found two little children seated at the foot of the hedge, crying aloud and clasping each other. At their feet a lamb was lying and bleating plaintively. And the younger was saying to the other: "Joseph, get up; we will go home." The elder tried to rise, but his legs, seized with a convulsive trembling, could not support him. "Joseph, Joseph, speak to me," said the poor little one; "speak to me." And Joseph, his teeth chattering, looked at his brother with eyes so big they frightened him. "There is one more apple in the basket; would you like it? I will give you all of it," went on the little fellow, his cheeks bathed in tears. And the elder trembled and then became rigid, by fits and starts, and stared fixedly with eyes growing ever larger and larger.

It was then that Simon passed. He put the two children on the donkey, took the basket, and, followed by the lamb, hastened to the village.

When the unhappy mother saw Joseph, her dear Joseph, so well a few hours before, so rejoiced at taking his brother for a walk, and now unconscious, dying, it was a scene to melt the heart. "My God, my God!" cried she, crazed with grief, "take me and leave my son! Oh, my Joseph! Oh, my poor Joseph!"

And, covering him with kisses, she burst into cries of despair.

The doctor was summoned; the basket in which were still some of the black berries mistaken for cherries explained to him the cause of the sad event. "Deadly nightshade, great God!" he exclaimed under his breath. "Alas! It is too late." Broken-hearted, he ordered a potion, the efficacy of which he could not count on, for the poison had made irreparable progress. And, in fact, an hour later, while the mother, on her knees at the foot of the bed, was praying and weeping, a little hand was stretched out from under the coverings and placed all cold in hers. It was the last good-by: Joseph was dead.

The next day they buried the poor little one. The whole village attended the funeral. Emile and Jules returned from the cemetery so sad that for several days they did not think of asking their uncle the cause of this lamentable accident.

Since then, in the house of mourning, little Louis stops playing every now and then and begins to cry, despite his beautiful tin watch. He has been told that Joseph has gone far away and that he will come back some day. "Mother," he says sometimes, "when will Joseph come back? I am tired of playing alone." His mother kisses him and, covering her face with a corner of her apron, sheds hot tears. "Don't you love Joseph any more, and is that why you cry when I speak of him?" asks the poor little innocent. And his mother, overwhelmed, tries in vain to stifle her sobs.

CHAPTER LVIII

POISONOUS PLANTS

THE DEATH of poor Joseph had spread consternation through the village. If children left the house and went off into the fields, there was constant anxiety until they returned. They might find poisonous plants that would tempt them with their flowers or their berries, and poison them. Many said, with reason, that the best way to prevent these terrible accidents was to know the dangerous plants and teach the children to beware of them. They went and found Maître Paul, whose great knowledge was appreciated by all, and asked him to teach them the poisonous plants of the neighborhood. So Sunday evening there was a numerous gathering at Uncle Paul's. Besides his two nephews and his niece, Jacques and Mother Ambroisine, there were Simon, who had come upon the two unfortunate children on his way home from the mill, Jean the miller, André the plowman, Philippe the vine-dresser, Antoine, Mathieu, and many others. The day before, Uncle Paul had taken a walk in the country to gather the plants he was to talk about. A large bunch of the principal poisonous plants, some in blossom, others with berries, were in a pitcher of water on the table.

"There are people, my friends," he began, "who shut their eyes so as not to see danger, and think themselves safe because they wilfully ignore peril. There are others who inform themselves about what may be a menace to them, persuaded that one warned person may be worth two unwarned. You belong to this latter class, and I congratulate you. Countless ills lie in wait for us; let us try to diminish their number by our vigilance, instead of giving ourselves up to lazy carelessness. Now that a frightful misfortune has overtaken one of our families, who does not realize the extreme importance of our all knowing, so as to avoid them, these terrible plants that claim victims every year? If this knowledge were more extended, the poor little fellow whose loss we now lament would still be his mother's consolation. Ah! unfortunate child!"

Uncle Paul, whom thunder never caused even to knit his brows, had tears in his eyes and his voice trembled. The good Simon, who had seen the two children in each other's arms under the hedge, felt more moved than the others at this recollection. He pulled down the broad rim of his hat to hide the big tears that were rolling down his rough cheeks bronzed by the sun. After a few moments of silence Uncle Paul continued:

"The death of the unfortunate little boy was caused by belladonna.

BELLADONNA

It is a rather large weed with reddish bell-shaped flowers. The berries are round, purplish-black, and resemble cherries. The leaves are oval and pointed at the end. The whole plant has a nauseous odor and a somber appearance, as if to announce the poison it conceals. The berries particularly are dangerous because they may tempt children by their resemblance to cherries and their sweetish taste. Enlargement of the pupil of the eye and a dull, fixed stare are the characteristics of belladonna poisoning."

Paul took from the bouquet in the pitcher a sprig of belladonna, and passed it around in the audience so that each one could examine the plant closely.

"What do you say that is called?" asked Jean.

"Belladonna."

"Belladonna; good. I know that weed. I have often found it near the mill, in shady places. Who would believe those pretty cherries held such a frightful poison."

Here André asked: "What does the word belladonna mean?"

"It is an Italian word meaning fine lady. Formerly, it seems, ladies used the juice of this plant to keep their complexion white."

"That is a property that does not concern our brown skin. What concerns us is this confounded berry which may tempt our children."

"Are not our herds in danger when this weed grows in pastures?" Antoine next inquired.

"It is very seldom that animals touch poisonous plants; they avoid browsing what might harm them, warned by the odor, and above all by instinct.

"This other plant with large leaves, whose flowers, red on the outside and spotted on the inside with white and purple, are arranged in a long and magnificent cluster almost as high as a man, is called digitalis. The flowers have the form of long, tun-bellied bells, or rather of glove-fingers; therefore it is called by different names, all referring to this peculiarity."

"If I am not mistaken," said Jean, "it is what we call fox-glove. It is common on the edges of woods."

"We call it fox-glove on account of its resemblance to the thumb of a glove. For the same reason it has elsewhere the name of gloves of Notre-Dame, gloves of the Virgin, and finger-stall. The name digitalis, borrowed from the Latin, also refers to the finger-shaped flower."

FOX-GLOVE

"It is a great pity that fine plant is poisonous," commented Simon; "it would be a pleasure to see it in our gardens."

"It is, indeed, cultivated as an ornamental plant, but in gardens under stricter vigilance than ours. As for us, my friends, who hardly have time to watch over flowers, we shall do well not to put digitalis within reach of children by introducing it in our gardens. The whole plant is poisonous. It has the singular property of slowing up the beating of the heart and finally stopping it. It is unnecessary to tell you that when the heart no longer beats, all is over.

"Hemlock is still more dangerous. Its finely-divided leaves

resemble those of chervil and parsley. This resemblance has often occasioned fatal mistakes, all the easier, because the formidable plant grows in the hedges of enclosures and even in our gardens. A plain enough characteristic, however, enables us to distinguish the poisonous weed from the two pot-herds that resemble it: that is the odor. Rub that tuft of hemlock in your hands, Simon, and smell."

"Ouf!" said Simon, "that smells very bad; parsley and chervil have not that horrid odor. When one is warned, no mistake can be made, in my opinion."

HEMLOCK

"Yes, when one is warned; but those who are not take no account of the smell and mistake hemlock for parsley or chervil. It is in order to be warned that you are listening to me this evening."

"You are doing us a great service, Maître Paul," said Jean, "by putting us on our guard against these dangerous plants. Every one at home ought to know what you have just taught us, so as not to gather a salad of hemlock instead of chervil."

ARUM

"There are two kinds of hemlock. One, called the great hemlock, is found in damp and uncultivated places. It is very like chervil. Its stems are marked with black or reddish spots. The other, called the little hemlock, resembles parsley. It grows in cultivated fields, hedges, and gardens. Both have a nauseating odor.

"Now here is a poisonous plant very easy to recognize. It is the arum, or, as it is commonly called, cuckoopint or calves'-foot. The arum is common in hedges. The leaves are very broad and shaped like a large lance-head. The blossom is shaped like a donkey's ear. It is a large yellowish trumpet, from the bottom of which rises a fleshy rod that might be taken for a little finger of butter. This strange flower is succeeded by a bunch of berries as large as peas and of a splendid red color. The whole plant has an unbearable burning taste."

"Let me tell you, Maître Paul," put in Mathieu, "what happened one day to my little Lucien. Coming home from school, he saw in the hedge those large flowers you are speaking of, like donkey's ears; the fleshy rod in the middle looked to him like something good to eat. You have just compared it to a little finger of butter. The thoughtless creature was taken with its looks. He bit into the deceitful finger of butter. What had he done! In a moment his tongue began to burn as if he had bitten a red-hot coal. I saw him come home spitting and making faces. He won't be taken in again, you may be sure. Luckily he hadn't swallowed the piece. The next morning he was all right."

ACONITES

"A similar burning flavor is found in the white milk-like juice that runs from the euphorbia when cut. The euphorbia are plants of mean appearance, very common everywhere. Their flowers, small and yellowish, grow in a head, the even branches of which radiate at the top of the stem. These plants are easily recognized by their white juice, their milk, which runs in abundance from the cut stems. This juice is dangerous, even on the skin alone, if it is tender; its acrid, burning taste is its sufficient characteristic.

"The aconites, like digitalis, are fine plants which for their beauty have been introduced in gardens, notwithstanding the violence of their poison. They are found in hilly countries. Their blossoms are blue or yellow, helmet-shaped, and grow in an elegant terminal bunch of the finest effect. Their leaves, of a lustrous green, are cut out in radiating sprays. The aconites are very poisonous. The violence of their poison has given them the name of dog's-bane and wolf's-bane. History tells us that formerly arrow-heads and lance-heads were soaked in the juice of the aconites, to poison the wounds made in war and to make them mortal.

"There is sometimes cultivated in our gardens a shrub with large shiny leaves, which do not fall in winter, and with black, oval berries as large as acorns. It is the cherry-bay. All its parts, leaves, flowers, and berries, have the odor of bitter almonds and peach kernels. The leaves of the cherry-bay are sometimes used to give their perfume to cream and milk products. They should be used only with great prudence, for the cherry-bay is extremely poisonous. They even say one has only to remain some time in its shade to become indisposed from its exhalations of a bitter-almond odor.

"In autumn there is seen in abundance, in damp fields, a large and beautiful flower, rose or lilac in color, that rises from the ground alone, without stem or leaves. It is the colchicum, called also meadow saffron, or veillotte, also veilleuse, because it blossoms on the eve of the cold season. If you dig a little way down, you will find that this flower starts from a rather large bulb, covered with a brown skin. Colchicum is poisonous; so cows never touch it. Its bulb is still more poisonous.

"But we have talked enough about harmful plants for to-day. I should be afraid of befogging your memories were I to enter into more details. Next Sunday I will expect you again, my friends, and will talk to you about mushrooms."

CHAPTER LIX

THE BLOSSOM

Y ES, THEY had listened very attentively the day before when Uncle Paul told them all about poisonous plants. Who would not listen to a talk on flowers? Jules and Claire, however, would have been glad to hear more. How are the flowers made that their uncle showed them yesterday? What is to be seen inside them? Of what use are they to the plant? Under the big elder-tree in the garden their uncle talked to them as follows:

"Let us begin with the blossoms of the digitalis, which I spoke of yesterday. Here is one. It has, as you see, almost the form of a glove-finger, or better, of a long pointed cap. Emile could put one on to his little finger; there would be plenty of room. It is purplish-red in color. Inside, it has spots of dark red encircled with white. The red glove-finger rises from the center of a circle of five little leaves. These little leaves are also part of the flower. Together they form what is called the calyx. The rest, the red part, is called the corolla. Remember these words, which are new to you."

"The corolla is the colored part of the flower; the calyx is the circle of little leaves at the base of the corolla," repeated Jules.

"Most flowers have two envelopes like these, one within

MALLOW

the other. The exterior, or calyx, is nearly always green; the interior, or corolla, is embellished with those magnificent tints that please us in so many flowers.

"In the mallow, which you see here, the calyx consists of five little green leaves, and the corolla of five large pieces of lilac rose-color. Each of these pieces is called a petal. The petals, all together, make the corolla."

"The corolla of the digitalis has only one piece or petal; that of the mallow has five," remarked Claire.

"It looks that way at first, but on examining closely you will find that they both have five. I must tell you that in a great many flowers the petals unite as soon as they begin to form in the bud, and by their union constitute a corolla which looks like only one piece. But very often the united petals separate a little at the edge of the flower, and by indentations more or less deep show how many are joined together.

"Look at this tobacco blossom. The corolla forms a tun-bellied funnel, apparently composed of only one piece. But the edge of the flower is cut out in five similar parts, which are the extremities of so many petals. The tobacco blossom, then, has five petals, the same as the mallow; only, these five petals, instead of being separate all their length, are welded together in a sort of funnel.

"Corollas with separate petals are called polypetalous corollas."

TOBACCO

"Like that of the mallow," suggested Claire.

"And that of the pear, almond, and strawberry," added Jules.

"Jules forgets some very pretty ones: the pansy and violet," said Emile.

"Corollas with petals all joined together are called monopetalous corollas," continued Uncle Paul.

"For example, digitalis and tobacco," said Jules.

"And the bell-flowers, don't forget them, the beautiful white bell-flowers that climb the hedges," Emile added.

"The five petals joined together are just as easily distinguishable in this flower we have here, called snap-dragon."

"Why is it called snap-dragon?" asked Emile.

"Because when it is pressed on both sides it opens its mouth like an animal."

Uncle Paul made the flower yawn; under pressure of his fingers it opened and shut its mouth as if biting. Emile looked on in amazement.

"In this mouth there are two lips, upper and lower. Well, the upper lip is split in two by a deep indentation, the sign of two petals, and the lower lip is split in three, indicating three petals. The corolla of the snap-dragon, although apparently all in one piece, is therefore in reality composed of five petals welded together."

SNAP-DRAGON

"There are, then," said Claire, "five petals in the mallow, pear, almond, digitalis, tobacco, and snap-dragon, with this difference, that the five petals are separate in the mallow, pear, and almond, and welded together in the digitalis, snap-dragon, and tobacco."

"Five petals, either separate or united," Uncle Paul went on, "are found in a great many other flowers.

"Let us come back to the calyx. The little green leaves of which it is composed are called sepals. There are five in the different flowers we have just examined, five in the mallow, five in tobacco, five in digitalis, five in the snap-dragon. Like the petals, the parts of the calyx, or sepals, sometimes remain separate, sometimes join together, but generally leave some indentations showing their number.

"The calyx having its parts distinct from one another is called a polysepalous calyx. That of the digitalis and of the snap-dragon is of this class.

"The calyx with sepals united is known as a monosepalous calyx. Such is that of the tobacco blossom. By the five indentations at its edge one can easily see that it is formed of five pieces joined together."

"The number five occurs again and again," observed Claire.

"A flower, my child, is beyond doubt a wonderful thing of beauty, but especially is it a masterpiece of wise construction. Everything about it is calculated according to fixed rules, everything arranged by number and measure. One of the most frequent arrangements is in sets of five. That is why we have just found five petals and five sepals in all the flowers examined this morning.

"Another grouping that often occurs is that in threes. It is found in bulb flowers,—the tulip, lily, lily of the valley, etc. These flowers have no green covering or calyx; they have only a corolla composed of six petals, three in an inner circle, three in an outer.

"The calyx and the corolla are the flower's clothing, a double clothing having both the substantial material that guards from inclemency, and the fine texture that charms the eye. The calyx, the outer garment, is of simple form, modest coloring, firm structure, suitable for withstanding bad weather. It has to protect the flower not yet opened, to shield it from the sun, from cold, and wet. Examine the bud of a rose or mallow; see with what minute precision the five sepals of the calyx are united to cover the rest. Not the slightest drop of water could penetrate the interior, so carefully are their edges joined together. There are flowers that close the calyx every evening as a safeguard against the cold.

"The corolla or inner garment unites elegance of form and richness of tint with fineness of texture. It is to the flower what wedding garments are to us. That is what especially captivates our eye, so that we commonly consider it the most essential part of the flower, while it is really only a simple ornamental accessory.

"Of the two garments, the calyx is the more necessary. Many flowers, of severe taste, know how to dispense with the pleasing part, the corolla; but they are very careful not to renounce the useful, the calyx, which, in its simplest form, is reduced to a tiny little leaf like a scale. Flowers without corolla remain unseen, and the plants that bear them seem to us to have no blossoms. It is a mistake: all trees and plants bloom."

"Even the willow, oak, poplar, pine, beech, wheat, and so many others whose blossoms I have never seen?" asked Jules.

"Even the willow, oak, and all the others. Their blossoms are extremely numerous, but, as they are very small and have no corolla, they escape the inattentive eye. There is no exception: every plant has its blossom."

CHAPTER LX

FRUIT

"IT WOULD be knowing a person very little only to be aware of his wearing a garment of a certain material, a coat of such and such a cloth. One does not know a flower any better when one knows that it is clothed with a calyx and a corolla. What is under this covering?

"Let us examine together this gillyflower. It has a calyx of four sepals and a corolla of four yellow petals. I take away these eight pieces. What is left now is the essential part; that is to say, the thing without which the flower could not fill its rôle and would be perfectly useless. Let us go carefully over this remaining part. You will find it well worth the trouble.

"First, there are six little white rods, each one surmounted by a bag full of yellow powder. These six pieces are called stamens. They

A FLOWERING BRANCH OF
THE GILLYFLOWER

are found in all flowers in greater or less number. The gillyflower has six, four longer ones arranged in pairs, and two shorter.

"The double bag that surmounts the stamen is called an anther. The dust contained in the anther is known as pollen. It is yellow in the gillyflower, lily, and most plants; ashy gray in the poppy."

"You have already told us," Jules interposed, "how clouds of pollen, raised by the wind in the woods, are the cause of sup-

posed showers of sulphur that frighten people so."

"I take away the six stamens. There remains a central body, swollen at the bottom, narrow at the top, and surmounted by a kind of head wet with a sticky moisture. In its entirety this central body takes the name of pistil; the swelling at the bottom is called an ovary, and the sticky head that terminates it is a stigma."

"What big names for such little things!" exclaimed Jules.

"Little, yes; but of unparalleled importance. These little things, my dear friend, give us our daily bread; without the miraculous work of these little things we should die of hunger."

"I will take care to remember their names, then."

"I, too," chimed in Emile; "but you must go over them again, they are so hard to learn."

Uncle Paul began again. Jules and Emile repeated after him: stamen, anther and pollen; pistil, stigma and ovary.

"With a penknife I divide the flower in two. The split ovary shows us what is inside."

"I see little seeds in regular rows in two compartments," observed Jules.

"Do you know what those hardly visible seeds are?"

"Not yet."

"They are the future seeds of the plant. The ovary, then, is the part of the plant where the seeds form. At a certain time the flower withers; the petals wilt and fall; the calyx does the same, or remains to play the part of protector a while longer; the dried stamens break off; only the ovary remains, growing larger, ripening, and finally becoming the fruit.

"Every fruit—the pear, apple, apricot, peach, walnut, cherry, melon, strawberry, almond, chestnut—began by being a little swelling of the pistil; all these excellent things that the plant furnishes us for food were first ovaries."

"A pear began by being the ovary of a pear blossom?"

"Yes, my child; pears, apples, cherries, apricots, begin by being the ovaries of their respective flowers. I will show you an apricot in its blossom."

Uncle Paul took an apricot blossom, opened it with his penknife, and showed the children what is here shown in the picture.

"In the heart of the flower you see the pistil surrounded by numerous stamens. The head that terminates it at the top is the stigma; the swelling at the bottom is the ovary or future apricot."

"That little green thing would have been an apricot, full of sweet juice, that I like so much?" inquired Emile.

"That little green thing would have become an apricot like those Emile is so fond of. Now would you like to see the ovary that gives us bread?"

"Oh, yes! All these things are very curious," replied Jules.

"Better than that, very important."

WHEAT

Claire gave her uncle a needle at his request; then with the delicate patience necessary for this operation he isolated one of the numerous flowers of which the whole forms the ear of wheat. The delicate little flower displayed clearly, on the point of the needle, the different parts composing it.

"The blessed plant that gives us bread has not time to think of its toilet. It has such weighty things to attend to: it must feed the world! So you see what quiet clothes it wears! Two poor scales serve it for calyx and corolla. You can easily recognize three hanging stamens with their double sachets for anthers. The principal body of the flower is the tun-bellied ovary, which, when ripe, will be a grain of wheat. It is surmounted by the stigma, fashioned like a double plume of exquisite delicacy. Salute it, my children: behold the modest little flower that gives life to us all!"

CHAPTER LXI

POLLEN

"I N A few days, even in a few hours, a flower withers. Pistils, stamens, calyx, fade and die. Only one thing survives: the ovary, which will become fruit.

"Now, in order to outlive the other parts of the flower and remain on its stem when all the rest dries up and falls, the ovary, at the moment when blossoming is at its greatest vigor, receives a supplement of strength, I should almost say a new life. The magnificence of the corolla, its sumptuous colorings, its perfumes, serve to celebrate the solemn moment when this new vitality comes to the ovary. This great act accomplished, the flower has had its day.

"Well, it is the pollen, the yellow dust of the stamens, that gives this increase of energy without which the nascent seeds would perish in the ovary, itself withered. It falls from the stamens on to the stigma, always coated with a stickiness apt to hold it; and from the stigma, it makes its mysterious action felt in the depths of the ovary. Animated with new life, the nascent seeds develop rapidly, while the ovary swells so as to give them necessary room. The final result of this incomprehensible travail is the fruit, with its contents of seeds ready to germinate and produce new plants. Do not question me further about these wonderful things concerning which even the keenest observer ceases to see clearly. God only, the wisest of beings, knows how a grain of pollen can give birth to something that was not before, and can cause the ovary to feel the stirring

GRAINS OF
POLLEN

of the vital principle.

"I will tell you now how we know that the falling of the pollen on to the stigma is indispensable to the development of the ovary into fruit.

"Most flowers have both stamens and pistils. All those we have just looked at are in that class. But there are plants that have some flowers with stamens and others with pistils. Sometimes the flowers with stamens only and those with pistils only are found on the same plant; sometimes they are found on separate plants.

"Did I not fear to overcharge your memory, I would tell you that plants having flowers with stamens only and flowers with pistils only on the same plant are called monœcious plants. This expression means 'living in one house.' In a word, the flowers with stamens and those with pistils live together in the same house, since they are found on the same plant. The pumpkin, cucumber, melon, are monœcious plants.

"Vegetables whose flowers with stamens and flowers with pistils are found on different plants are termed diœcious; that is to say, plants with a double house. By this is meant that the ovary and pollen are not found in the same plant. The locust, date, and hemp are diœcious.

"The locust is a tree of extreme southern France. Its fruit grows in pods similar to those of the pea, but brown, very long, and plump. This fruit, in addition to seeds, has a sugary flesh. Supposing we took a notion, if the climate permitted, to grow locust seeds in our garden. What locust tree must we plant? Evidently the tree with pistils, because it alone possesses the ovaries which become fruit. But that is

FLOWERING BRANCH OF LOCUST TREE not enough. Planted by itself,

the locust tree with pistils will be able to blossom abundantly every year, without ever producing any fruit; for its flowers would fall without leaving a single ovary on the branches. What is wanting? The action of the pollen. Close to the locust with pistils let us plant one with stamens. Now fructification proceeds as we wish. Wind and insects carry the pollen from the stamens to the stigmas; the torpid ovaries spring to life, and in time the locust pods grow and ripen perfectly. With pollen, fruit; without pollen, no fruit. Are you convinced, Jules?"

"Without doubt, Uncle; only, unfortunately, we do not know the locust. I should prefer a plant of our own region."

"I will tell you of one that will permit you to prove what I have told you; but first of all let me mention a second example.

"The date-tree, like the locust, is diœcious. Arabs cultivate it for its fruit,—dates, their chief food."

"Dates are those long fruits of a very sweet taste, preserved dry in boxes," said Jules. "A Turk was selling some at the last fair. The kernel is long and split all along one side from one end to the other."

"That is it. In the country of

DATE-PALM

the date-tree, a sandy country burnt by the sun, spots of watered and fertile earth are rare. These spots are called oases. It is necessary to utilize them as much as possible. So the Arabs plant only date-trees with pistils, the only ones that will produce dates. But when they are in flower, the Arabs go long distances to seek bunches of flowers with stamens on wild date-trees, to shake the dust on the trees they have planted. Without this precaution there is no harvest."

"Uncle will tell us so much," Emile interposed, "that I shall have as much regard for the pollen as I have for the ovary.

Without it, I should not have tasted the dates of the Turk who smoked such a long pipe; without it, no apricots and no cherries."

"In the garden there is a long pumpkin-vine that will soon blossom. I will give it to you for the following experiment.

"The pumpkin is monœcious; flowers with stamens and flowers with pistils inhabit the same house, the same plant. Before they are full-blown they can easily be distinguished from each other. The flowers with pistils have under the corolla a swelling almost as large as a nut. This swelling is the ovary, the future pumpkin. The blossoms with stamens have not this swelling.

"Cut off all the blossoms with stamens before they are full-blown, and leave those with pistils. For greater surety, wrap each one of these in a piece of gauze before it is in full-bloom. The covering must be large enough to permit the flower to open. Do you know what will happen? Not being able to receive the pollen, since the flowers with stamens are cut off, and since, also, the gauze wrapping keeps out the insects from the neighboring gardens, the pistillate flowers will wither after languishing a while, and the plant will not produce any pumpkins.

"Would you, on the contrary, like such and such blossoms, at your choice, to produce pumpkins in spite of their gauze prison and the suppression of the staminate blossoms? With the tip of your finger take a little pollen from one of the blossoms you have cut off, and put the yellow dust on the stigma of a pistillate flower. Then replace the gauze wrapping. That is enough, the pumpkin will come."

"You will let us try that delightful experiment?" asked Jules.

"I will, I give the pumpkin-vine over to you."

"I have some gauze," volunteered Claire.

"And I some string to tie it with," added Emile.

"Come along," cried Jules.

And, gay as larks, the three children ran to the garden to get everything ready for the experiment.

CHAPTER LXII

THE BUMBLE-BEE

T HE FLOWERS with pollen were cut off, those with ovaries wrapped each in a separate gauze-bag. Every morning they went and watched the blossoming. With pollen taken from the cut flowers they powdered the stigmas of four or five pistillate blossoms. And it happened just as their uncle had said. The ovaries whose stigmas had received the pollen became pumpkins, the others dried up without swelling. Now, during these experiments, which were both a serious study and a joyful amusement, Uncle Paul continued his account of the flower.

"The pollen reaches the stigma in divers ways. Sometimes the stamens, which are longer, let it fall by its own weight on the shorter pistil. Sometimes the wind, shaking the flower, deposits the dust of the stamens on the stigma, or even carries it long distances for the benefit of other ovaries.

"There are flowers whose stamens behave in such a manner as to fulfil their mission. They bend over alternately and apply their anthers to the stigma, there to deposit some pollen; then slowly raise themselves to give place one to another. They might be regarded as a circle of courtiers depositing their offerings at the feet of a great king. These salutations at an end, the rôle of the stamens is finished. The flower fades,

DIŒCIOUS PLANTS (MALE AND FE-MALE) OF VALLISNERIA SPIRALIS

but the ovary begins to ripen its seeds.

"The vallisneria is a plant that lives under the water. It is very common in the Southern Canal. Its leaves resemble narrow green ribbons. It is diœcious, that is to say it has flowers with stamens and those with pistils on different plants. The pistillate flowers are borne on long, tightly curled stems. The blossoms with stamens have only very short stems. Under water, where the current would carry away the pollen and prevent its fastening itself on the stigmas, the quickening action of the stamens on the pistil cannot take place. So the vallisneria, fixed by its roots in the mud, is obliged to send its flowers to the surface of the water to let them blossom in the open air. It is easy for the pistillate flowers. They unwind the curl that supports them, and mount to the surface. But what will the staminate flowers do, fastened as they are to the bottom with their short stems?"

"I cannot undertake to say," answered Jules.

"Well, by their own strength, without any external aid, these flowers pull away from their stems, break their moorings, and mount to the surface to rejoin the pistillate flowers. Then they open their little white corollas and free their pollen to wind and insects, which deposit it on the stigmas. After that they die and the current carries them away, while the flowers quickened by the pollen curl up again and descend once more beneath the water, there to ripen their ovaries at leisure."

"It is wonderful, Uncle; one would say those little flowers know what they are doing."

"They do not know what they are doing; they obey mechanically the laws of Providence, which makes sport of difficulties and knows how to accomplish miracles in a simple blade of grass. Would you like another striking example of this infinite wisdom that foresees everything, arranges everything? Let us come back to the snap-dragon.

"Insects are the flower's auxiliaries. Flies, wasps, honeybees, bumble-bees, beetles, butterflies, all vie with one another in rendering aid by carrying the pollen of the stamens to the stigmas. They dive into the flower, enticed by a honeyed drop expressly prepared at the bottom of the corolla. In their efforts to obtain it they shake the stamens and daub themselves with pollen, which they carry from one flower to another. Who has not seen bumble-bees coming out of the bosom of the flowers all covered with pollen? Their hairy stomachs, powdered with pollen, have only to touch a stigma in passing to communicate life to it. When in the spring you see on a blooming pear-tree, a whole swarm of flies, bees, and butterflies, hurrying, humming, and fluttering, it is a triple feast, my friends: a feast for the insect that pilfers in the depth of the flowers; a feast for the tree whose ovaries are quickened by all these merry little people; and a feast for man, to whom abundant harvest is promised. The insect is the best distributor of pollen. All the flowers it visits receive their share of quickening dust."

"It is in order to prevent the insects coming from neighboring gardens and bringing pollen that you have had the pumpkin blossoms covered with bags of gauze?" inquired Emile.

"Yes, my child. Without this precaution the pumpkin experiment would certainly not succeed; for insects come from a distance, very far perhaps, and deposit on our flowers the pollen gathered from other pumpkins. And very little of it is necessary; a few grains are enough to give life to an ovary.

"To attract the insect that it needs, every flower has at the bottom of its corolla a drop of sweet liquor called nectar. From this liquor bees make their honey. To draw it from corollas shaped like a deep funnel, butterflies have a long trumpet, curled in a spiral when at rest, but which they unroll and plunge into the flower like a bore when they wish to obtain the delicious drink. The insect does not see this drop of nectar; however, it knows that it is there and finds it without hesitation. But in some flowers a grave difficulty presents itself: these flowers are closed tight everywhere. How is the treasure to be got at, how find the entrance that leads to the nectar? Well, these closed flowers have a signboard, a mark that says clearly: Enter here."

"You won't make us believe that!" cried Claire.

"I am not going to make you believe anything, my dear child; I am going to show you. Look at this snap-dragon blossom. It is shut tight, its two closed lips leave no passage between. Its color is a uniform purplish red; but there, just in the middle of the lower lip, is a large spot of bright yellow. This spot, so appropriate for catching the eye, is the mark, the signboard I told you of. By its brightness it says: Here is the keyhole.

BUMBLE-BEE

"Press your little finger on the spot. You see. The flower yawns immediately, the secret lock works. And you think the bumble-bee does not know these things? Watch it in the garden and you will see how it can read the signs of the flowers. When it visits a snap-dragon, it always alights on the yellow spot and nowhere else. The door opens, it enters. It twists and turns in the corolla and covers itself with pollen, with which it daubs the stigma. Having drunk the drop, it goes off to other flowers, forcing the opening of which it knows the secret thoroughly.

"All closed flowers have, like the snap-dragon, a conspicuous point, a spot of bright color, a sign that shows the insect the entrance to the corolla and says to it: Here it is. Finally, insects whose trade it is to visit flowers and make the pollen fall from the stamens on to the stigma, have a wonderful knowledge of the significance of this spot. It is on it they use their strength to make the flower open.

"Let us recapitulate. Insects are necessary to flowers to bring pollen to the stigmas. A drop of nectar, distilled on purpose for this, attracts them to the bottom of the corolla; a bright spot shows them the road to follow. Either I am a triple idiot or we have here an admirable chain of facts. Later, my children, you will find only too many people saying: This world is the product of chance, no intelligence rules it, no Providence guides it. To those people, my friends, show the snap-dragon's yellow spot. If, less clear-sighted than the burly bumble-bee, they do not understand it, pity them: they have diseased brains."

CHAPTER LXIII

MUSHROOMS

WHILE THEY were talking about insects and flowers, time had slipped by until the Sunday arrived when Uncle Paul was to tell about mushrooms. The gathering was larger than the first time. The story of poisonous plants had been repeated in the village. Some people in a rut, content with their stupid ignorance, had said: "What is the use of it?" "The use!" replied the others; "it teaches one to beware of poisonous plants, so as not to die miserably like poor Joseph." But those in the rut had tossed their heads with a satisfied air. Nothing is so sufficient unto itself as folly. So only willing listeners came to Uncle Paul.

"Of all poisonous plants, my friends," he began, "mushrooms are the most formidable; and yet some furnish a delightful food capable of tempting the soberest."

"For my part," observed Simon, "I acknowledge, nothing is equal to a dish of mushrooms."

"Nobody will accuse you of gluttony, for, as I have just said, mushrooms can tempt the soberest. I do not wish to discourage their use. I know too well what a resource they are in the country; I simply propose to put you on your guard against the poisonous kinds."

"You are going to teach us to distinguish the good from the bad?" asked Mathieu.

"No; that is impossible for us."

"How impossible? Everybody knows that you can eat without fear mushrooms that grow at the foot of such and such a tree."

"Before answering that remark, I will address myself to you all and ask: Have you confidence in my word? Do you think that passing one's life in studying such things is more instructive than the hear-say of those who do not concern themselves with these matters?"

"You may speak, Maître Paul: we all have full confidence in your learning," Simon made answer for the company.

"Well, then, I repeat it in all conviction: it is impossible for us who are not specialists to distinguish an edible mushroom from a poisonous one, for none has a mark to say: This is eatable and this is not. Neither the nature of the ground, nor the trees at the foot of which they grow, nor their form, color, taste, smell, can teach us anything or enable us to distinguish at sight the harmless from the poisonous. I admit that a person who had passed long years studying mushrooms with the minute attention of a scientist would succeed in distinguishing pretty well the poisonous from the harmless, just as one acquires a knowledge of any other plant; but can we undertake such studies? Have we the time? We scarcely know a dozen weeds, and yet we would presume to pass judgment on the properties of mushrooms, so many in kind and resembling one another so closely?

MUSHROOMS

"I hasten to add that, in every locality, actual use has long since taught the people some kinds that they can eat without danger. It is a good thing to conform to this usage, which makes us profit by other people's experience—on condition, be it understood, that we acquaint ourselves with the kinds used. But that is not enough to keep us safe from all peril. It is so easy to make a mistake! And then, go to another place and you will come across other mushrooms which, while apparently of the same family as those you have known as eatable, will be dangerous. My rule of conduct is, you see, absolute: you must beware of all mushrooms; excess of prudence is necessary here."

"I admit with you," said Simon, "that it is impossible for us to distinguish at sight the eatable from the poisonous kinds; but there are ways of deciding the question."

"Tell us how."

"In the autumn we cut mushrooms in slices and dry them in the sun. They are excellent food for winter. The poisonous mushrooms rot without drying. The good ones keep."

"Wrong. All mushrooms, good or bad indifferently, keep or spoil according to their more or less advanced state and according to the weather at the time of preparation. This characteristic is of no value whatever."

"Worms attack good mushrooms," Antoine here interposed; "they do not attack bad ones, because they poison them."

"That characteristic is no better than the other one. Worms attack all old mushrooms, bad as well as good; for what would be death to us is harmless to them. Their stomach is made so that they can eat poison with impunity. Certain insects eat aconite, digitalis, belladonna; they feast on what would kill us."

"They say," remarked Jean, "that a piece of silver put in the pot when the mushrooms are cooking turns black if they are poisonous, and remains white if they are good."

"The saying is a foolish one, and to put it in practice a folly. Silver does not change color any more from bad than from good mushrooms."

"There is nothing to do, then, but give up mushrooms. That would be hard on me," said Simon.

"No, no; I promise you, on the contrary, that you will be able to use them more than you have done. The only thing is to proceed advisedly.

"What is poisonous in mushrooms is not the flesh, but the juice with which it is impregnated. Get rid of that juice, and the injurious properties will disappear immediately. This is accomplished by slicing and cooking the mushrooms, either dried or fresh, in boiling water with a handful of salt. They are then drained in a colander and washed two or three times in cold water. That done, they are prepared in any way one chooses.

"If, on the contrary, mushrooms are prepared without having first been cooked in boiling water, we expose ourselves to the danger of a poisonous juice.

"The cooking in boiling water to which salt has been added is so efficacious that, in order to solve this serious problem, certain persons have had the courage to eat for whole months the most poisonous mushrooms, prepared, however, in the way I have just told you."

"And what happened to them?" asked Simon.

"Nothing at all. It is true that these persons prepared their poisonous mushrooms with the most scrupulous care."

"There was reason for it. According to you, then, one could use all mushrooms without distinction?"

POISONOUS MUSHROOM

"Strictly speaking, yes. But that would be going too far, much too far. There would be the fear of incomplete preparation, insufficient cooking. I only affirm that you must submit mushrooms of good repute in the neighborhood to the preliminary cooking in boiling water. If, by chance, some poisonous ones were included, the poison would in this way be eliminated and no accident would happen; I would bet my hand on that."

"What you have just taught us, Maître Paul, will be profited by, you may be sure. Are we ever quite certain that there is nothing poisonous in what we gather?"

Before saying good-by Simon approached Mother Ambroisine and entered with her into more circumstantial details of the cooking. He is so fond of mushrooms, the worthy man!

CHAPTER LXIV

IN THE WOODS

THE HISTORY of mushrooms reduced to a rule for cooking which will save us from grave dangers was enough for Simon, Mathieu, Jean, and the others, who lacked time to hear more; but Emile, Jules, and Claire were not satisfied: they wished to extend their knowledge on these strange vegetables. So their uncle took them one day to a beech wood near the village.

The trees, several hundred years old and with their branches meeting at a great height, formed an arch of foliage through which, here and there, shone a ray of sunlight. Their smooth trunks, with white bark, gave the effect of enormous columns sustaining the weight of an immense building full of shade and silence. On the lofty summits crows cawed while smoothing their feathers. Occasionally a redheaded green woodpecker, surprised at its work, which consists of pecking the wormy wood with its beak to make the insects come out that it feeds on, gave a cry of alarm and flew off like a dart. In the midst of the moss with which the ground was carpeted were here and there numbers of mushrooms. Some were round, smooth, and white. Jules could not admire them enough; he likened them in his imagination to eggs laid in a mossy hollow by some wandering hen. Others were glossy red, others bright fawn-color, and still others brilliant yellow. Some, just coming out of the ground, were enveloped in a kind of bag that tears open as the mushroom grows; some, more advanced, spread out like an open umbrella. Finally, there were many that had already begun to decay. In their fetid rottenness swarmed innumerable grubs, which later would become insects. After picking a number of the principal kinds, the party sat down at the foot of a beech, on the soft moss-carpet, and Uncle Paul spoke thus:

"A mushroom is the blossom of a plant that lives under ground and is called by learned men mycelium. This subterranean plant is composed of white, slender, fragile threads, resembling in their entirety a large cobweb. If you pull up a mushroom carefully you will see at the base of its stalk, in the earth that clings to it, numerous white threads of the mycelium. Let us imagine a rosebush planted so as to leave nothing but the roses above ground. The buried bush will represent the subterranean mycelium; the roses, open to the air, will represent the blossoms of the mycelium, that is to say the mushrooms."

"A rosebush," objected Jules, "has stout branches covered with leaves; the mushroom-plant, according to what I see, has nothing of the sort. It is a kind of moldiness that branches out in the ground in white veins."

"Those white veins, so delicate that one can hardly touch them without breaking them, form the subterranean plant, without leaves or roots. They lengthen little by little in the ground to a pretty good distance from the point of departure. Then, at a favorable moment, they produce little swellings which grow under ground, become mushrooms, and burst open their bed of earth to expand in the air. This structure explains to us why mushrooms grow in groups. Each group, with the mycelium that produces it, constitutes one and the same plant."

"I have seen groups of mushrooms in a perfect circle," Claire remarked.

"If the ground is of uniform character and nowhere hinders the propagation of the subterranean vegetable in one direction rather than in another, the mycelium spreads equally on all sides, and so produces circular groups of mushrooms, which the country people sometimes call witches' circles."

"Why witches' circles?" asked Jules.

"The ignorant and superstitious think they see an effect of witchcraft in this curious circular arrangement, whereas it is but the natural result of the uniformly equal development of the subterranean plant."

"Then there are no witches?" said Emile.

"No, my dear. There are rogues who abuse the credulity of

others; there are simpletons disposed to listen to them; but no one has preternatural powers."

"Since a mushroom is the blossom of a subterranean plant, of the mycelium, as you call it, must it not have stamens, pistils, ovaries?" Jules inquired.

"A mushroom is in its way the blossom of a kind of vegetable, but its structure has nothing in common with that of ordinary flowers. It is a structure of a special sort, very complicated, very curious, which I shall pass by in silence so as not to overcharge your memory.

"The chief function of a flower, you know, is to produce seeds. Well, the mushroom too produces seeds, but so small, so different from others, that they have a special name,— spores. Spores are the seed of the mushroom, just as acorns are the

MUSHROOMS

seed of the oak. That is worthy of some further explanation.

"The mushrooms most familiar to us are composed of a sort of dome supported by a stalk. This dome is called the cap. The under side of the cap takes various shapes, of which the principal are these: Sometimes it is composed of gills which radiate from the center to the border; sometimes it is pierced by an infinity of little holes, which are the orifices of as many tubes joined together in a common mass; sometimes it is covered with fine points like those of a cat's tongue.

"Mushrooms that have the under side of the cap formed of radiating gills are called agarics; those pierced with little holes, boleti; those covered with little points, hydnei. Agarics and boleti are the most common."

Here Uncle Paul took, one by one, the mushrooms they had gathered and showed his nephews the gills of the agarics, the holes of the boleti, and the points of the hydnei.

CHAPTER LXV

THE ORANGE-AGARIC

"**M**USHROOM SEEDS, or spores, form on these gills, these points, and on the walls of the tubes of which these holes are the orifices. I recommend to Jules the following experiment. We will take some mushrooms whose caps are not yet thoroughly spread. We will place them this evening on a sheet of white paper. During the night the blossoming will be finished and the ripe seeds will fall from the gills of the agarics and the tubes of the boleti. To-morrow morning we shall find on the paper an impalpable dust, red, rose, brown, according to the kind of mushroom.

"This dust is nothing but a mass of seeds, of spores, so fine that they cannot be seen separately without a microscope, so numerous they cannot be counted. There are millions and millions of them."

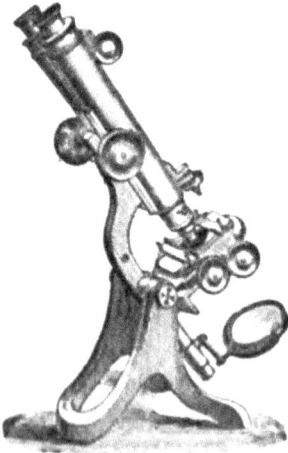

"A microscope," interrupted Emile. "Is that the instrument with which you sometimes look at things so small that the naked eye can scarcely see them?"

"Yes. A microscope enlarges the objects seen through it, and shows them to us in all their details of structure, although they would be hidden from the unaided eye by their smallness."

"Will you show us through the microscope the mushroom spores when I have collected them on a sheet of paper?" asked Jules.

BINOCULAR MICROSCOPE

"I will show them to you. One spore is enough, under favorable

conditions of heat and moisture, to germinate and develop into white filaments or
mycelium from which will spring at the
right time numerous mushrooms. How
many mushrooms would be produced
if all the spores that fall by myriads and
myriads from the gills of a single agaric
were to germinate? Here again we have
the story of the cod, the louse, all the
feeble creatures, in short, that reproduce
their kind in such immense numbers."

"To have mushrooms, then, as many
as we want, it is only necessary to sow
the spores?" Jules again inquired.

"In that you are mistaken, my dear
child. Up to this time mushroom culture

SPORES

has been impossible, because the care required by their excessively delicate seeds is not understood by us, or may even be
beyond our power. Only one edible mushroom is cultivated, and
even in growing this we use not the spores, but the mycelium.

"They call it the hot-bed mushroom. It is an agaric, satiny
white above and pale rose beneath. In the old stone quarries
near Paris they make beds of horse manure and light earth. In
these beds they put pieces of mycelium known to horticultur-
ists under the name of mushroom-spawn. This spawn ramifies,
pushes out numerous filaments, and from these finally spring
the mushrooms."

"Good to eat!"

"Excellent. Among the mushrooms we gathered are those
that I am going to acquaint you with.

"Look at this first of all. It is an agaric. The upper surface of
the cap is a beautiful orange-red; the gills underneath are yellow. The stalk rises from the bottom of a sort of white bag with
torn edges. This bag, called volva at first enveloped the whole
mushroom. In growing and pushing above ground, the cap broke
it. This kind, they say, is the best of all, the most appreciated. It
is called the orange-agaric.

"This other agaric, likewise orange-red, and also provided

with a bag or volva at the base of the stalk, is called the false orange-agaric. Would you not, however, think it was the same kind?"

"I don't see much difference, for my part," responded Claire.

"Nor I either," said Emile.

"I see a difference," Jules declared, "but it is very slight. The second agaric has white gills, while the first has yellow."

"Jules has sharp eyes. I will add that in the false orange-agaric the upper surface of the cap is sown with shreds of white skin, debris of the torn volva. The other one has not these shreds, or very few.

"If one did not pay attention to these slight differences, one would commit a very fatal error. The first mushroom is a delicious viand; the second, or false orange-agaric, is a deadly poison."

"I am no longer surprised," said Jules, "at your telling Simon that it is impossible for us, without long study, to distinguish the good from the bad kinds. Here are two mushrooms almost as much alike as two drops of water: one kills, the other is excellent."

"Not a year passes without its lamentable cases of poisoning, from a confusion of the two kinds. Remember carefully their characteristics, so as not to expose yourself some day to a terrible mistake."

"I will be very careful not to forget them," Jules promised. "Both orange-agarics are orange-red and have a white volva or bag. The eatable orange-agaric has yellow gills; the poisonous one, white gills."

"Besides," added Emile, "the poisonous orange-agaric has numerous shreds of white skin on the cap."

"Look at this other that I picked from the trunk of a tree. It is a large, dark-red boletus. It has no stalk. It fastens itself to old trunks by one of its sides. It is called the tinder-agaric boletus, because its flesh, cut in thin slices, dried in the sun, and made flexible by hammering, makes tinder."

"I did not dream that tinder came from a mushroom." said Jules.

"The truffle is the most important of eatable mushrooms. It grows under ground, like the mycelium that produces it. Its odor betrays its presence. A very keen-scented animal, the pig, is led into the wood. Enticed by the smell of the subterranean mushroom, the pig roots with its snout at the spots where the truffles are hidden. Then the pig is driven away, but to console him they throw him a chestnut; and finally the precious mushroom is dug up. In its shape the truffle bears no resemblance to ordinary mushrooms. It has a bulky round body, wrinkled, and black flesh marbled with white."

CHAPTER LXVI

EARTHQUAKES

E ARLY IN the morning all the neighbors were talking, from door to door, on the same subject. It seemed they had had a narrow escape during the night. Jacques said that about two o'clock he had been awakened by the bellowing of his cattle, repeated two or three times. Even Azor himself, the good Azor, so peaceful in his stall when there was nothing serious to disturb him, had bellowed mournfully. Jacques had risen and lighted his lantern, but had been unable to discover what caused the trouble with the animals.

Mother Ambroisine, who slept with one eye open, told a longer tale. She had heard the dishes rattling on the kitchen dresser; some plates had even rolled off and broken in falling to the ground. Mother Ambroisine was thinking it was perhaps some misdeed of the cat's, when it seemed to her that strong arms seized the bed and shook it twice from head to foot and from foot to head. It was over in the twinkling of an eye. The worthy woman was so frightened that, throwing the covers over her head, she commended her soul to God.

Mathieu and his son were away at the time: they were returning home from the fair, and were making the journey by night. The weather was fine—no wind, and bright moonlight. They were chatting about their affairs when a dull, deep noise was heard, coming from under the ground. It sounded like the roar of the big mill-dam. At the same moment they staggered as if the ground had been giving way under them. Then nothing more. The moon continued to shine, the night was calm and serene. It was so soon over that Mathieu and his son wondered whether they had not dreamed it.

These were among the more serious incidents related. Meanwhile there was passing from mouth to mouth, moving some to incredulous smiles and others to grave reflections, the terrible word "earthquake."

In the evening Uncle Paul was surrounded by his auditors, eager for some explanation of the great news of the day.

"Is it true, Uncle," asked Jules, "that the earth sometimes trembles?"

"Nothing is truer, my dear child. Sometimes here, sometimes elsewhere, suddenly there is a movement of the ground. In our privileged countries we are far from having any exact idea of these terrible agitations of the earth. If once in a while a slight trembling is felt, it is talked of for days as a curiosity; then it is forgotten. Many tell to-day of the events of the past night without attaching much importance to them, not knowing that the force revealed to us by a light movement of the earth can, in its brutal power, bring about frightful disasters. Jacques has told you of the bellowing of the cattle and Azor's outcry. Mother Ambroisine has described to you her fright when her bed was shaken twice. In all that there is nothing very terrifying; but earthquakes are not always harmless. Alas, no; and may God preserve us from ever undergoing the sad experience!"

"Is an earthquake, then, very serious?" Jules again inquired. "For my part, I thought it only meant a few plates broken and some furniture displaced."

"It seems to me," said Claire, "that if the movement were strong enough houses would fall down. But Uncle is going to tell us about a violent earthquake."

"Earthquakes are often preceded by subterranean noises, a dull rumbling that swells, abates, swells again, as if a storm were bursting in the depths of the earth. At this rumbling, full of menacing mysteries, every creature becomes quiet, mute with fear, and every one turns pale. Warned by instinct, animals are struck with stupor. Suddenly the earth shivers, bulges up, subsides again, whirls, cracks open, and discloses a yawning gulf."

"Oh, my goodness!" Claire exclaimed. "And what becomes of the people?"

"You will see what becomes of them in these terrible catastrophes. Of all the earthquakes felt in Europe, the most terrible was that which ravaged Lisbon in 1775, on All Saints' Day. No danger appeared to menace the festal town, when suddenly there burst from under-ground a rumbling like continuous thunder. Then the ground, shaken violently several times, rose up, sank down, and in a moment the populous capital of Portugal was nothing but a heap of ruins and dead bodies. The people that were still left, seeking refuge from the fall of the ruins, had retired to a large quay on the seashore. All at once the quay was swallowed up in the waters, dragging with it the crowd and the boats and ships moored there. Not a victim, not a piece of wreck came back to float on the surface. An abyss had opened, swallowing up waters, quay, ships, people, and, closing up again, kept them for ever. In six minutes sixty thousand persons perished.

"While that was happening at Lisbon and the high mountains of Portugal were shaking on their bases, several towns of Africa—Morocco, Fez, Mequinez—were overthrown. A village of ten thousand souls was swallowed up with its entire population in an abyss suddenly opened and suddenly closed."

"Never, Uncle, have I heard of such terrible things," declared Jules.

"And I laughed," said Emile, "when Mother Ambroisine told us of her fright. It was nothing to laugh at. If it had been God's will, our village might last night have disappeared from the earth with us all, as did that one in Africa."

"Listen to this, too," Uncle Paul continued. "In February, 1783, in Southern Italy, convulsions began that lasted four years. During the first year alone nine hundred and forty-nine were counted. The surface of the ground was wrinkled in moving waves like the surface of a stormy sea, and on this unstable ground people felt nauseated as if on the deck of a vessel. Sea-sickness reigned on land. At every undulation, the clouds, really immobile, seemed to move bruskly, just as they do at sea when we are on a vessel tossed by the winds. Trees bowed in the terrestrial wave and swept the earth with their tops.

"In two minutes the first shock overthrew the greater part of towns, villages, and small boroughs of Southern Italy, as well as of Sicily. The whole surface of the country was thrown into confusion. In several places the ground was creviced with fissures, resembling on a large scale the cracks in a pane of broken glass. Vast tracts of ground, with their cultivated fields, their dwellings, vines, olive-trees, slid down the mountain-sides and went considerable distances, to settle finally on other sites. Here, hills split in two; there, they were torn from their places and transported to some other part. Elsewhere, there was nothing to uphold the ground, and it was engulfed in yawning abysses, taking with it dwellings, trees, and animals, which were never seen again; in still other places, deep funnels full of moving sand opened, forming presently vast cavities that were soon converted into lakes by the inrush of subterranean waters. It is estimated that more than two hundred lakes, ponds, and marshes were thus suddenly produced.

"In certain places the ground, softened by waters turned from their channels or brought from the interior by the crevices, was converted into torrents of mud that covered the plains or filled the valleys. The tops of trees and the roofs of ruined farm buildings were the only things to be seen above this sea of mud.

"At intervals sudden quakes shook the ground to a great depth. The shocks were so violent that street pavements were torn from their beds and leaped into the air. The masonry of wells flew out from below the surface in one piece, like a small tower thrown up from the earth. When the ground rose and split open, houses, people, and animals were instantly swallowed up; then, the ground subsiding again, the crevice closed once more, and, without leaving a vestige, everything disappeared, crushed between the two walls of the abyss as they drew together. Some time afterward, when, after the disaster, excavations were made in order to recover valuable lost objects, the workmen observed that the buried buildings and all that they contained were one compact mass, so violent had been the pressure of this sort of vise formed by the two edges of the closed-up crevice.

"The number of persons who perished in these terrible circumstances is estimated at eighty thousand.

"Most of these victims were buried alive under the ruins of their houses; others were consumed by fires that sprang up in these ruins after each shock; others, fleeing across the country, were swallowed up in the abysses that opened under their feet.

"The sight of such calamities ought to have awakened pity in the hearts of barbarians. And yet—who would believe it?—except for a very few acts of heroism, the conduct of the people was most infamous. The Calabrian peasants ran to the towns, not to give help, but to pillage. Without any concern about the danger, they traversed the streets in the midst of burning walls and clouds of dust, kicking and robbing the victims even before the breath had left their bodies."

"Miserable creatures!" cried Jules. "Horrid rascals! Ah, if I had only been there!"

"If you had been there, what would you have done, my poor child? There were plenty there with as good hearts and better fists than yours, but they could do nothing."

"Are those Calabrians very wicked?" asked Emile.

"Wherever education has not been introduced there are brutal natures that, in time of trouble, spring up, no one knows whence, and frighten the world with their atrocities. Another story will teach you more of the Calabrian peasants."

CHAPTER LXVII

SHALL WE KILL THEM BOTH?

U NCLE PAUL went up to his room and came back with a book. "What I am going to read to you is from a mounted artilleryman, more expert in the art of the pen than in that of the cannon. At the beginning of this century a French army occupied Calabria. Our gunner belonged to it. Here is a letter he wrote to his cousin:

"'One day I was traveling in Calabria. It is a country of bad people who love no one and have a special spite against the French. It would take too long to tell you why; enough that they mortally hate us and one is sure of a bad time if one falls into their hands.

"'My companion was a young man. In these mountains the roads are precipices; our horses could hardly climb them. My comrade was in front. A path that seemed to him shorter and more practicable misled us. It was my fault. Ought I to have put my trust in a man of twenty years? As long as daylight lasted we tried to find our way through the woods; but the more we tried the more bewildered we got, and it was pitch dark when we reached a dimly lighted house. We entered, not without suspicion, but what could we do?

"'There we found a charcoal-burner and all his family at table, to which they immediately invited us. My young man needed no urging. We sat down, eating and drinking, or he at least, for I busied myself examining the place and the countenances of our hosts. They had the appearance of charcoal-burners, but the house might have been taken for an arsenal. It was full of guns, pistols, sabers, knives, cutlasses. It all displeased me, and I saw well that I on my part was equally displeasing to our entertainers.

"'My comrade, on the contrary, made himself one of the family; he laughed, chaffed with them, and, with an imprudence that I ought to have foreseen, told them at the very first whence we came, whither we were going, who we were. Frenchmen, imagine it! Amongst our most mortal enemies, alone, lost, far from all human aid; and then, to add to our probable ruin, he acted the rich man, promising these people whatever they wished in payment and for the hire of guides on the morrow. Finally, he spoke of his valise, begging them to be very careful of it and to put it at the head of his bed: he said he did not wish any other bolster. Ah! youth, youth, how your immaturity is to be pitied! Cousin, you would have thought we were carrying the crown diamonds!'"

"That young man was certainly very imprudent," commented Jules. "Could he not hold his tongue, seeing he was in the hands of wicked people?"

"Silence is very difficult for giddy, careless young persons. I will go on:

"'Supper finished, they left us. Our hosts slept below, we in the upper room where we had eaten. A loft seven or eight feet high, reached by a ladder, was the bed that awaited us—a kind of nest that one got into by crawling under joists laden with provisions for a year. My comrade climbed up alone and was soon asleep, his head on the precious valise; I determined to watch, so made a good fire and sat down by it.

"'The night had almost passed, quietly enough, and I began to feel reassured, when, just as it seemed to me it must be near daylight, I heard our host and his wife quarreling immediately under me, and, putting my ear close to the fire-place that communicated with the one below, I distinguished perfectly this proposal of the husband: "Well, now, let us see; shall we kill them both?" To which the woman answered: "Yes." And I heard nothing more.

"'What can I say? I remained scarcely breathing, my body cold as marble. God! When I think of it! We two all but unarmed against those twelve or fifteen with so many weapons! And my comrade dead with sleep and fatigue! To make a noise by calling him, I dared not; to escape by myself, I could not. The window was not far from the ground, but beneath it two big dogs were howling like wolves.'"

"Poor gunner!" Emile exclaimed.

"And his comrade sleeping like a simpleton!" Claire added.

"'At the end of a quarter of an hour, which seemed long, I heard some one on the stairs, and through the cracks of the door I saw the father, a lamp in one hand and one of his large knives in the other. He was coming up, his wife following him. I placed myself behind the door as he opened it; he put down the lamp, and his wife came and took it; then he entered, barefoot. From outside she said to him in a low tone, shading the lamp with her hand: "Gently, go gently!" When he came to the ladder, he mounted, knife between his teeth, and reaching the height of the bed on which lay this poor young man, his throat uncovered, with one hand he grasped his knife, and with the other—Ah! cousin—'"

"Enough, Uncle; this story frightens me!" cried Claire.

"Wait—'And with the other he seized a ham that was hanging from the ceiling, cut off a slice, and went off the way he had come. The door closed, the lamp disappeared, and I was left alone with my reflections.'"

"And then?" inquired Jules.

"And then, nothing more. 'As soon as it was daylight,' continued the gunner, 'the whole family came and awakened us with much noise, as we had requested them. They brought food and served us a very good breakfast, I assure you. Two capons were part of it, one of which our hostess said we must eat, and take the other with us. On seeing them I understood the significance of those terrible words: Shall we kill them both?'"

"The man and woman were discussing whether they should kill both capons or only one for breakfast?" asked Emile.

"That and nothing else," replied his uncle.

"All the same, the gunner had a bad quarter of an hour for his mistake."

"Those charcoal-burners were not at all such bad people as I thought at first," said Jules.

"That is the point I wished to make. Calabria, like all countries, has its good and its bad people."

CHAPTER LXVIII

THE THERMOMETER

"THE STORY of the gunner," Jules remarked, "ended very differently from what one expected at the beginning. Just when one thinks the two travelers are done for, it turns out nothing more serious is in question than the roasting of two fowls. A shiver of fear seizes you when the man climbs the ladder with the cutlass between his teeth; the next minute you are laughing. That is a very amusing story; but it has turned us aside from the earthquakes. You have not told us yet the cause of these terrible movements of the ground."

"If that interests you," replied his uncle, "let us talk about it a little. I will tell you first that the farther you descend into the earth, the hotter it becomes. Excavations made by man for obtaining various minerals give us valuable information on this subject. The deeper they go, the hotter it is. For every thirty meters of depth there is an increase of one degree in temperature."

"I don't know very well what a degree is," said Jules.

"And I don't know anything about it," confessed Emile.

"Let us begin with that; if not, it would be impossible for you to understand. In my room you have seen, on a little wooden board, a glass rod pierced by a very fine canal and ending at the bottom in a little bulb. In the bulb is a red liquid, which ascends or descends in the canal of the tube according to whether it is warmer or colder. That is called a thermometer. In freezing water the red liquid goes down to a point in the tube called zero; in boiling water it goes up to a point marked 100. The distance between these two points is divided into one hundred equal parts called degrees."[2]

---

2   It is the centigrade thermometer that is here described.—Translator.

"Why degrees?" asked Emile.

"By that it is meant that these divisions have a certain resemblance to the degrees or steps of a flight of stairs, or the rounds of a ladder. The red liquid goes up or down from division to division just as we mount or descend a flight of stairs step by step. If it grows warmer, the red liquid moves and little by little climbs the steps; if colder, it goes down the ladder. Thus the heat can be estimated according to the step or degree where the liquid stops.

"It is freezing when the liquid goes down to zero; the heat is that of boiling water when it goes up to division 100. The intermediate steps or degrees indicate, evidently, other states of heat, greater when the degree is higher up on the ladder.

"The degree of heat of any body, as indicated by the thermometer, is called its temperature. Thus we say the temperature of freezing water is zero, that of boiling water one hundred degrees."

"One morning," said Emile, "when you sent me to get something from your room, I put my hand on the little bulb of the thermometer. The red liquid began to go up, little by little."

"It was the warmth of your hand that made it go up."

"I wanted to see how high the liquid would go, but I had not patience to wait till the end."

"I will tell you. At last the thermometer would have marked at most 38 degrees, which is the temperature of the human body."

"And in the very hot days of summer what degree does the thermometer mark?" asked Jules.

"In our region the greatest heat of summer is from 25 to 35 degrees."

"And in the hottest countries of the world?" Claire inquired.

"In the hottest countries, Senegal, for example, the temperature rises to 45 and 50 degrees. It is twice as hot as our summer."

CHAPTER LXIX

THE SUBTERRANEAN FURNACE

"LET US get back to our subject. At the bottom of mines, I told you, a high temperature prevails, which keeps up during the whole year. There is always the same heat, winter and summer. The deepest excavation miners have ever made is in Bohemia. It is inaccessible to-day. Landslides have partly filled it. At the depth of 1151 meters the thermometer indicated a perpetual heat of forty degrees, almost the temperature of the hottest regions in the world. And that, mind you, in winter as well as summer. When mountainous Bohemia was covered with ice and snow, it was only necessary to go down to the bottom of the mine to pass from the rigors of winter to the insupportable heat of a Senegal summer. One shivered with cold at the entrance and stifled with heat at the bottom.

"The same conditions, without exception, prevail everywhere. The deeper one descends in the earth, the hotter one finds the temperature. In deep mines the heat is such that the most unobservant workman is struck by it and wonders if he is not near some immense furnace."

"The interior of the earth is, then, really a stove?" queried Jules.

"Much more than a stove, as you will see. The name of artesian well is given to a cylindrical hole which by means of strong iron bars, fitted end to end, is made in the ground until some reservoir of subterranean water, fed by the infiltrations of neighboring streams or lakes, is reached. The water that comes up from far under ground as the result of such a boring reaches the surface at a temperature equal to that of those depths; and thus we learn about the distribution of heat in the bowels of the earth. One of the most remarkable of these wells is that of Grenelle, at Paris. It is 547 meters deep, and the water in it is constantly at 28 degrees, a temperature almost as high as that of the hottest summer days. The water of the artesian well of Mondorf, on the frontier of France and Luxemburg, comes from a far greater depth, 700 meters. Its temperature is 35 degrees. Artesian wells, of which there are at present a considerable number, illustrate the same principle as mines: for every thirty meters of depth the heat increases one degree."

"Then by digging wells deep enough we should at last come to boiling water?"

"Certainly. The difficulty is to attain the desired depth. To reach the temperature of boiling water it would be necessary to bore about three quarters of a league, which is impossible. However, a number of natural springs are known which, as they come from the ground, possess a high temperature, sometimes reaching the boiling point. They are called thermal springs, which means hot springs. There prevails, then, at the depth from which they come, a heat sufficient to make them tepid, or even boiling hot. The most remarkable hot springs of France are those of Chaudes-Aigues and Vic, in Cantal. They are almost boiling."

"Do these springs make streams that are different from others?"

"Steaming streams, in which you can plunge an egg for a moment and take it out cooked."

"Then there are no little fish or crabs," said Emile.

"Certainly not, my dear. You understand that if there were any they would be cooked through and through."

"That is true."

"The little streams of boiling water in Auvergne are nothing in comparison with what are seen in Iceland, that large island situated at the extreme north of Europe and covered with snow the greater part of the year. It has numbers of springs throwing up hot water, called in that country geysers. The most powerful, or the Great Geyser, springs from a large basin situated on the top of a hill formed by the smooth white incrustations deposited by the foam of the water. The interior of this basin is funnel-shaped and terminates in tortuous conduits penetrating to unknown depths.

"Each eruption of this volcano of boiling water is announced by a trembling of the earth and dull noises like distant detonations of some subterranean artillery. Every moment the detonations become stronger; the earth trembles, and, from the bottom of the crater, the water rushes up in an impetuous torrent and fills the basin, where, for a few moments, we have what looks like a boiler heated by some invisible furnace. In the midst of a whirlpool of steam the water rises in a boiling flood. Suddenly the geyser musters all its force: there is a loud explosion, and a column of water six meters in diameter spouts upward to the height of sixty meters, and falls again in steaming showers after having expanded in the shape of an immense sheaf crowned with white vapor. This formidable outburst lasts only a few moments. Soon the liquid sheaf sinks; the water in the basin retires, to be swallowed up in the depths of the crater, and is replaced by a column of steam, furious and roaring, which spouts upward with thunderous reverberations and, in its indomitable force, hurls aloft huge masses of rock that have fallen into the crater, or breaks them into tiny bits. The whole neighborhood is veiled in these dense eddies of steam. Finally calm is restored and the fury of the geyser abates, but only to burst forth again later and repeat the same program."

Giant Geyser, Yellowstone National Park

"That must be terrible and beautiful at the same time," commented Emile. "No doubt you look at this furious fountain from a long distance, so as not to be struck on the back by boiling showers."

"What you have just told us, Uncle," said Jules, "shows plainly that there is great heat under ground."

"In admitting, as all these observations justify us in doing, that the subterranean temperature increases with the depth one degree for every thirty meters, it is estimated that at three kilometers or three quarters of a league down, the temperature must be that of boiling water, that is to say 100 degrees. Five leagues down, the heat is that of red-hot iron; at twelve leagues it is sufficient to melt all known substances. At a greater depth the temperature, apparently, is still higher. Accordingly we are to imagine the earth is formed of a globe of matter liquefied by fire and enveloped by a thin crust of solid material that is upborne by that central ocean of melted minerals."

"You say," said Claire, "a thin crust of solid material; and yet, according to the calculations you have just mentioned, the thickness of the solid material must be about twelve leagues. Under that would be the melted matter. It seems to me twelve leagues make a good thickness, and we have nothing to fear from the subterranean fire."

"Twelve leagues are very little in relation to the earth's dimensions. The distance from the surface of the earth to its center is 1600 leagues. Of this distance about twelve leagues belong to the thickness of the solid crust, all the rest to the molten globe. On a ball two meters in diameter the solid crust of the earth would be represented by a thickness of half a finger's breadth. Let us make a more simple comparison, representing the earth by an egg. Well, the egg-shell is the solid crust of the globe; its liquid content is the central mass in fusion."

"And we are separated from the immense subterranean furnace only by that thin shell!" exclaimed Jules. "That is not at all reassuring."

"I agree, it is not without a certain emotion that one hears for the first time what science tells us of these intimate details of the earth's structure; one cannot think without fear of those burning abysses that roll their waves of melted minerals a few leagues under our feet. How can a covering, relatively so light, resist the fluctuations of the central liquid mass? This fragile crust, this shell of the globe, will it not some time melt, become disjointed, crumble, or at least move? The little it does move makes continents tremble and the ground crack open in frightful chasms."

"Ah!" interposed Claire, "that is the cause of earthquakes. The liquid that is inside is stirred, and the shell moves."

"It seems to me," Jules remarked, "that this shell, comparatively so thin, ought to tremble oftener."

"Perhaps not a day passes without the solid crust of the earth experiencing some shock, sometimes at one point, sometimes at another, beneath the bed of the seas, as well as under the continents. However, disastrous earthquakes are very rare, thanks to the intervention of volcanoes.

"Volcanic orifices are, in fact, veritable safety-valves, which put the interior of the globe in communication with the exterior. By offering permanent vents to the subterranean vapors that tend to liberate themselves by overturning the earth, they render earthquakes less frequent and less disastrous. In volcanic countries every time the ground is shaken by strong shocks, the earthquake ceases the moment the volcano begins to throw up its fumes and lava."

"I well remember," said Jules, "your account of the eruption of Etna and the Catanian disaster. At first I only saw in volcanoes terrible mountains spreading devastation around them; now I begin to see their great use, their necessity. Without their air-holes, the earth would seldom be still."

CHAPTER LXX

SHELLS

Iₙ Uɴᴄʟᴇ Paul's room was a drawer full of shells of all sorts. One of his friends had collected them in his travels. Pleasant hours could be passed in looking at them. Their beautiful colors, their pleasing but sometimes odd shapes, diverted the eye. Some were twisted like a spiral stair-case, others widened out in large horns, still others opened and closed like a snuff-box. Some were ornamented with radiating ribs, knotty creases, or plates laid one on another like the slates of a roof; some bristled with points, spines, or jagged scales. Here were some smooth as eggs, sometimes white, sometimes spotted with red; others, near the rose-tinted opening, had long points resembling wide-stretched fingers. They came from all parts of the world. This came from the land of the negroes, that from the Red Sea, others from China, India, Japan. Truly, many pleasant hours could be passed in examining them one by one, especially if Uncle Paul were to tell you about them.

One day Uncle Paul gave his nephews this pleasure: he spread before them the riches of his drawer. Jules and Claire looked at them with amazement; Emile was never tired of putting the large shells to his ear and listening to the continual hoo-hoo-hoo that escapes from their depths and seems to repeat the murmur of the sea.

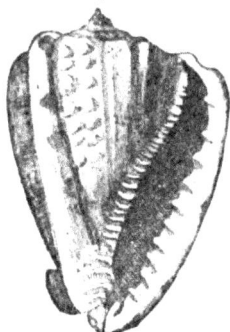

"This one with the red and lace-like opening comes from India. It is called a helmet. Some are so large that two of them would be as much as Emile could

Cᴀssɪs

245

carry. In some islands they are so abundant that they are used instead of stones and are burnt in kilns to make lime."

"I would not burn them for lime," said Jules, "if I found such beautiful shells. See how red the opening is, how beautifully the edges are pleated."

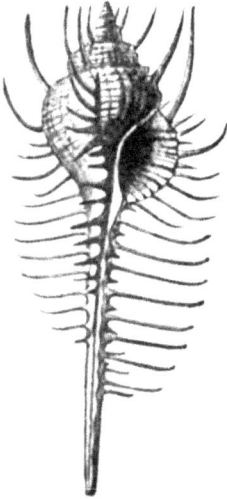

SPINY MOLLUSK

"And then what a loud murmur it makes," added Emile. "Is it true, Uncle, that it is the noise of the sea echoed by the shell?"

"I do not deny that it resembles a little the murmur of waves heard at a distance; but you must not think that the shell keeps in its folds an echo of the noise of the waves. It is simply the effect of the air going and coming through the tortuous cavity.

"This other belongs to France. It is common on the shores of the Mediterranean and belongs to the genus cassis."

"It goes hoo-hoo, like the helmet," Emile remarked.

"All those that are rather large and have a spiral cavity do the same.

"Here is another which, like the preceding, is found in the Mediterranean. It is the spiny mollusk. The creature that inhabits it produces a violet glair, from which the ancients derived, for their costly stuffs, a magnificent color called purple."

"How are shells made?" asked Claire.

"Shells are the dwellings of creatures called mollusks, the same as the spiral snail's shell is the house of the horny little animal that eats your young flowering plants."

"Then the snail's house is a shell, the same as the beautiful ones you have shown us," Jules observed.

"Yes, my child. It is in the sea that we find, in greatest number, the largest and most beautiful shells. They are called sea-

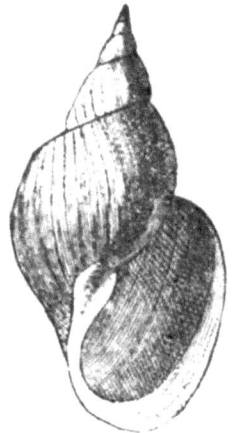

PALUDINIDÆ

shells. To these belong the helmet-shell, cassidula, and spiny mollusk. But fresh waters, that is to say streams, rivers, ponds, lakes, have them too. The smallest ditch in our country has shells of good shape but somber, earthy in color. They are called fresh-water shells."

"I have seen some in the water resembling large, pointed, spiral snails," said Jules. "They have a sort of cap to close the opening."

"They are Paludinidæ."

"I remember another ditch shell," said Claire. "It is round, flat, and as large as a ten or even twenty-sou piece."

"That is one of the Planorbinæ. Finally, there are shells that are always found on land and for that reason are called land-shells. Such is the spiral snail."

"I have seen very pretty snails," Jules remarked, "almost as pretty as the shells

PLANORBINÆ

in this drawer. In the woods you see yellow ones with several black bands wound round them in regular order."

"The creature we call the spiral snail—isn't it a slug that finds an empty shell and lives in it?" asked Emile.

"No, my friend; a slug remains always a slug without becoming a snail; that is to say, it never has a shell. The snail, on the contrary, is born with a tiny shell that grows little by little as the snail grows. The empty shells you find in the country have had their inhabitants, which are now dead and turned to dust, only their houses remaining."

"A slug and a snail without its shell are very much alike."

"Both are mollusks. There are mollusks that do not make shells, the slug for example; others that do make them, such as the snails, the Paludinidæ, and the cassididæ."

"And of what does the snail make its house?"

"Of its own substance, my little friend; it sweats the materials for its house."

"I don't understand."

"Don't you make your teeth, so white, shiny, and all in a row? From time to time a new one pushes through, without your giving it any thought. It does it by itself. These beautiful teeth are of very hard stone. Where does that stone come from? From your own substance, it is clear. Our gums sweat stone which fashions itself into teeth. So the snail's house is built. The little creature sweats the stone that shapes itself into a graceful shell."

"But to arrange stones one on another and make houses of them you need masons. The snail's house is made without masons."

"When I say it is done by itself, I do not mean that the stone has the faculty of making itself into a shell. You never see rubble piling itself unaided into a wall. God, the Father of all things, willed that the stone should arrange itself in a mother-of-pearl palace to serve as a dwelling for the poor animal, brother to the slug, and it is accomplished according to His will. In like manner He told the stone to grow up into beautiful teeth from the depths of the rosy gums of little boys and girls, and it is done as He willed."

"I begin to feel rather friendly toward the snail, the voracious animal that eats our flowers," said Jules.

"I do not care to make you friendly with it. Let us make war on it since it ravages our gardens; it is our right; but do not let us disdain to learn from it, for it has many beautiful things to teach us. To-day I will tell you of its eyes and nose."

CHAPTER LXXI

THE SPIRAL SNAIL

"WHEN THE snail crawls, it bears aloft, as you know, four horns."

"Horns that come out and go in at will," added Jules.

"Horns that the animal turns every way," said Emile, "when you put the shell on the live coals. Then the snail sings be-be-be-eou-eou."

"Stop that cruel play, my child. The snail does not sing; it is complaining, in its own way, of the fiery tortures. Its slime, coagulated by the heat, first swells and then shrinks, and the air that escapes by little puffs produces that dying wail.

"In one of La Fontaine's fables, where there are so many good things about animals, he tells us that the lion, wounded by a horned animal,

"Straight banished from his realm, 't is said,
All sorts of beasts with horns—
Rams, bulls, goats, stags, and unicorns.
Such brutes all promptly fled.
A hare, the shadow of his ears perceiving,
Could hardly help believing
That some vile spy for horns would take them,
And food for accusation make them.
Adieu, said he, my neighbor cricket;
I take my foreign ticket.
My ears, should I stay here,
Will turn to horns, I fear;
And were they shorter than a bird's,
I fear the effect of words.

These horns! the cricket answered; why,
God made them ears; who can deny?
Yes, said the coward, still they'll make them horns,
And horns, perhaps, of unicorns!
In vain shall I protest.[3]

"This hare evidently exaggerated things. Its ears have remained ears, to all observers. We do not know whether the snail
exiled himself in these circumstances; man is almost unanimous
in regarding as horns what the snail bears on its forehead. 'You
call those horns!' the cricket would have exclaimed, being better
advised than man; 'you must take me for a fool.'"

"Then they are not horns?" asked Jules.

"No, my dear. They are at once hands, eyes, nose, and a cane
for the blind. They are called tentacles. There are two pairs of
unequal length. The upper pair is the longer and more remarkable.

"Right at the end of each long tentacle you see a little black
point. It is an eye as complete as that of the horse and ox, in
spite of its minute dimensions. What is necessary for making
an eye, you are far from suspecting. It is so complicated I will
not try to tell you. And yet it is all to be found in that little black
point that is scarcely visible. That is not all: beside the eye is a
nose, that is to say an organ especially sensitive to odors. The
snail sees and smells with the tips of its long tentacles."

"I have noticed that if you bring anything near the snail's

ELEPHANT

3   The translation is that of Elizur Wright, Jr., published by James Miller,
New York, 1879.

long horns, the animal draws them in."

"This combination of nose and eye can retreat, advance, go to meet an object, and catch odors from all sides. To find a similar nose, you must go from a snail to an elephant, whose trunk is an exceptionally long nose. But how much superior the snail's is to the elephant's! Sensitive to odors and light, eye and nose at the same time, it can retire within itself like the finger of a glove, disappear by reëntering the animal's body, or come out from under the skin and lengthen itself like a telescope."

"I have often seen how the snail pulls his horns in," observed Emile. "They fold back inward and seem to bury themselves under the skin. When anything annoys it, the animal puts its nose and eyes into its pocket."

"Precisely. To protect ourselves from too strong a light or an unpleasant odor, we shut our pupils and stop up our nose. The snail, if the light troubles or some smell displeases it, sheathes eyes and nose in their covering; it puts them into its pocket, as Emile says."

"It is an ingenious way," Claire remarked.

"You said, too," interposed Jules, "that the horns served it as a blindman's cane."

"The animal is blind when it has drawn in its upper tentacles, partly or wholly; it then has only the two lower ones, which explore objects by the touch better than does the cane of a blind man, for they are very sensitive. The two upper tentacles, besides their functions of eye and nose, also play the part of blindman's cane, or, better still, that of a finger that touches and recognizes objects. You see, little Emile, one does not know everything about a snail when one knows its wail on the fire."

"I see. Who of us would have suspected that those horns are eyes, nose, blindman's cane, fingers, all at the same time?"

CHAPTER LXXII

MOTHER-OF-PEARL AND PEARLS

"SOME OF the shells you have just shown us," said Jules, "shine inside like the handle of that pretty penknife you bought me the day of the fair—you know?—that four-bladed penknife with the mother-of-pearl handle."

"That is plain enough. Mother-of-pearl, that pretty substance that shines with all the colors of the rainbow, comes from certain shells. We use for delicate ornamentation what was once the dwelling of a glairy animal, near relation to the oyster. Truly, this dwelling is a veritable palace in richness. It shines with all imaginable tints, as if the rainbow had deposited its colors there.

"This is the shell that furnishes the most beautiful mother-of-pearl. It is called the meleagrina margaritifera. Outside it is wrinkled and blackish-green; inside it is smoother than polished marble, richer in color than the rainbow. All tints are found there, bright, but soft and changeable, according to the point of view."

"That superb shell is the house of a miserable, slimy animal! In fairy tales the fairies themselves have none to equal it. Oh! how beautiful, how beautiful it is!"

"Every one has his portion in this world. The slimy animal has for his a splendid palace of mother-of-pearl."

"Where does the meleagrina live?"

"In the seas that wash the shores of Arabia."

"Is Arabia very far away?" in-

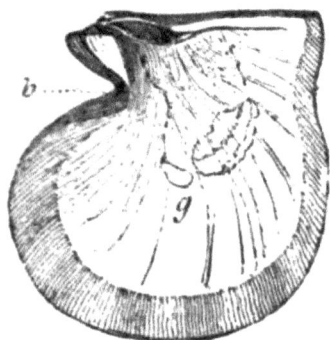

MELEAGRINA (AVICULA)
MARGARITIFERA

quired Emile.

"Very far, my dear. Why do you ask?"

"Because I should like to pick up a lot of these beautiful shells."

"Don't dream of such a thing. It is too far away, and, besides, they are not to be gathered by every one that wants them. To get the meleagrina men have to dive to the bottom of the sea, and some of them never come up again."

"And there are people who dare to dive to the bottom of the sea just to get shells?" asked Claire.

"Plenty. So profitable, too, is the trade that we should be badly received by the first-comers if we took a notion to go and fish with them."

"Then those shells are very precious?"

"You shall judge for yourself. First the inner layer of the shell, sawed into sheets and tablets, is the mother-of-pearl that we use for fine ornamentation. Jules' penknife-handle is covered with a sheet of mother-of-pearl that was part of the inside of a pearl-shell. But that is the least part of what the precious shell produces. There are pearls as well."

"But pearls are not very dear. With a few sous I bought a whole boxful, to embroider you a purse."

"Let us make a distinction: there are pearls and pearls. The pearls you mention are little pieces of colored glass pierced with a hole. Their price is very moderate. The pearls of the meleagrina are globules of the richest and finest mother-of-pearl. If they are unusually large, they attain the fabulous price of the diamond, up to hundreds of thousands and millions of francs."

OYSTER SHELL

"I don't know those pearls."

"God keep you from ever knowing them, for in becoming interested in pearls one sometimes loses common sense and honor. It is well, though, to know how they are produced.

"Between the two parts of the shell lives an animal like the oyster. It is a mass of slime in which you would find it difficult to recognize an animal. It digests, however, and breathes, and is sensitive to pain, so sensitive that a grain of dust, a mere nothing, renders existence painful to it. What does the animal do when it feels itself tickled by some foreign substance? It begins to sweat mother-of-pearl around the place that itches. This mother-of-pearl piles up in a little smooth ball, and there you have a pearl made by the sick, slimy animal. If it is of any considerable size, it will cost a fine bag of crowns, and the person who wears it around her neck will be very proud of it.

"But before getting to the neck, it must be fished for. The fishermen are in a boat. They descend into the sea, one after another, with the aid of a rope to which is tied a large stone that drags them rapidly to the bottom. The man about to dive seizes the weighted rope with his right hand and the toes of his right foot; with his left hand he closes his nostrils; to his left foot is fastened a bag-shaped net. The stone is thrown into the sea. The man sinks like lead. Hastily he fills the net with shells, and then pulls the rope to give the signal for ascent. Those in the boat pull him up. Half-suffocated, the diver reaches the surface with his fishing. The efforts he has made to suspend respiration are so painful that sometimes blood gushes from his mouth and nose. Sometimes, the diver comes up with a leg gone; sometimes he never comes up. A shark has swallowed him.

SHARK

"Some of those pearls that shine in a jeweler's windows cost much more than a fine bag of crowns: they may have cost a man's life."

"If Arabia were at the end of the village, I would not go pearl-fishing," declared Emile.

"To open the shells, they are exposed to the sun until the animals are dead. Then men rummage in the pile, which smells horribly, and get the pearls. There is nothing more to do except pierce them with a hole."

"One day," said Jules, "when they were cleaning the big mill-race I found some shells that shone inside like mother-of-pearl."

"We have in our streams and ditches shells in two parts of a greenish black. They are called fresh-water mussels. Their inside is mother-of-pearl. Some, very large and living by preference in mountain streams, even produce pearls. But these pearls are far from having the luster and consequently the price of those of the meleagrina."

CHAPTER LXXIII

THE SEA

"Do all those beautiful shells you have in the drawer come from the sea?" asked Emile.

"They come from the sea."

"Is the sea very large?"

"So large that in certain parts it takes ships whole months to go from shore to shore. They are fast vessels, too, especially the steamships. They go almost as fast as a locomotive."

"And what is to be seen at sea?"

"Overhead, the sky as here; all around, a large, blue, circular expanse, and beyond that nothing. One travels leagues and leagues, and yet is always in the middle of that blue circle of waters, as if one had not advanced. The rounded form of the earth, and consequently of the seas covering the greater part of it, is the cause of this appearance. The eye can take in only a small extent of the sea, an extent bounded by a circular line on which the dome of the sky appears to rest; and as the circle of the waters is ever being renewed while keeping the same appearance as one advances, it seems as if one remained stationary in the center of the circle where the blue of the sky merges into the blue of the sea. However, by dint of this continued advance one finally perceives a little gray smoke on the line that bounds the view. It is land beginning to show. Another half-day's journey, and the little gray smoke will have become rocks on the coast or high mountains in the interior."

"The sea is larger than the earth, the geography says," remarked Jules.

"If you divide the surface of the terrestrial globe into four equal parts, land will occupy about one of these parts, and the sea, taken all together, the other three."

"What is under the sea?"

"Under the sea there is ground, the same as under the waters of a lake or stream. Under-sea ground is uneven, just as dry land is uneven. In certain parts it is hollowed out into deep chasms that can scarcely be sounded; in others it is cut up with mountain-chains, the highest points of which come up above the level of the water and form islands; in still others, it extends in vast plains or rises up in plateaus. If dry, it would not differ from the continents."

"Then the depth is not the same everywhere?"

"In no wise. To measure the depth of the water, a plummet attached to one end of a very long cord is cast into the sea; the length of line unrolled by the plummet in its fall indicates the depth of the water.

"The greatest depth of the Mediterranean appears to be between Africa and Greece. In these parts, in order to touch bottom, the lead unwinds 4000 or 5000 meters of line. This depth equals the height of Mont Blanc, the highest mountain in Europe."

"So if Mont Blanc were set down in this hollow," was Claire's comment, "its summit would only just reach the surface of the water."

"There are deeper places than that. In the Atlantic, south of the banks of Newfoundland, one of the best spots for cod-fishing, the lead shows about 8000 meters. The highest mountains in the world, in Central Asia, are 8840 meters high."

"Those mountains would come up above the surface of the water in the place you spoke of, and would form islands 850 meters in height."

"Finally, in the seas about the South Pole there are places where the lead shows 14,000 or 15,000 meters of depth, or nearly 4 leagues. Nowhere has the dry land any such altitudes.

"Between these fearful chasms and the shore where the water is no deeper than the thickness of one's finger, all the intermediate degrees may be found, sometimes varying gradually, sometimes suddenly, according to the configuration of the ground underneath. On one shore the sea increases in depth with frightful rapidity. That shore is, then, the top of an escarpment of which the sea washes the base. On another it increases little by little, and one must go a long distance to attain a depth of a few meters. There the ocean bed is a plain, sloping almost imperceptibly, in continuation of the terrestrial plain.

"The average depth of the ocean appears to be from six to seven kilometers; that is to say, if all the submarine irregularities were to disappear and give place to a level bed, like the bottom of a basin made by man, the seas, while preserving on the surface their present extent, would have a uniform layer of water of from 6000 to 7000 meters in depth."

"I get rather bewildered with all these kilometers," complained Emile. "Never mind; I begin to understand that there is a great deal of water in the sea."

"Much more than you could ever imagine. You know the Rhone, the largest river in France; you have seen it at flood, when its muddy waters form a sheet from one bank to the other as far as the eye can reach. It is estimated that in this condition it pours into the sea about five million liters of water a second. Well, if it always preserved that majestic fulness, this large river could not, in twenty years, fill the thousandth part of the ocean basin. Does that make you understand any better how immense the sea is?"

"My poor head is dizzy at the mere thought of it. What color are the waters of the sea? Are they yellow and muddy like the Rhone?"

"Never, except at the mouths of rivers. Seen in a small quantity, the water looks colorless; seen in a great mass, it appears of its natural color, greenish blue. The sea, then, is blue with a greenish tinge, darker in the open sea, clearer near the coasts. But this coloring changes a great deal, according to the brightness of the sky. Under a bright sun the calm sea is now pale blue, now dark indigo; under a stormy sky it becomes bottle-green and almost black."

"The waters of the great deep"

CHAPTER LXXIV

WAVES SALT SEAWEEDS

"WHERE DO the waves come from?" asked Jules. "The sea is very terrible, they say, when it is angry."

"Yes, my dear Jules, very terrible. I shall never forget those great moving ridges, capped with foam, that toss a heavy ship like a nutshell, carry it one moment on their monstrous backs, then let it plunge into the liquid valley that intervenes. Oh! how small and weak one feels on those four planks, mounting and plunging at the will of the waves! If the nutshell springs a leak under the furious blows of the billows, may the good God have pity on us! The shattered boat would disappear in fathomless depths."

"In the chasm you told us about?" Claire asked.

"In those chasms from which no one returns. The shattered boat would be swallowed up in the sea, and nothing of you would be left but a remembrance, if there were people left on the earth who loved you."

"So the sea ought always to be calm," said Jules.

"It would be a pity, my child, if the sea were always at rest. This calm would be incompatible with the salubrity of the seas, which must be violently stirred up to keep them free from taint and to dissolve the air necessary to their animal and vegetable population. For the ocean of waters, as for the atmosphere or ocean of air, there is need of a salutary agitation—of tempests that churn up, renew, and vivify the waters.

"The wind disturbs the surface of the ocean. If it comes in gusts, it creates waves that leap with foaming crest and break against one another. If it is strong and continuous, it chases the waters in long swells, in waves or surges that advance from the open in parallel lines, succeed one another with a majestic uniformity, and one after another rush booming on to the shore. These movements, however tumultuous they may be, affect only the surface of the sea; thirty meters down the water is calm, even in the most violent storms.

"In our seas the height of the biggest waves is not more than two or three meters; but in some parts of the South Sea the waves, in exceptional weather, rise to ten or twelve meters. They are veritable chains of moving hills with broad and deep valleys between. Whipped by the wind, their summits throw up clouds of foam and roll up in formidable volume with a force sufficient to shatter the largest vessels under their weight.

"The power of the waves borders on the prodigious. There, where the shore, rising vertically from the water, presents itself fully to the assaults of the sea, the shock is so violent that the earth trembles under one's feet. The most solid dikes are demolished and swept away; enormous blocks are torn off, dragged along the ground, sometimes thrown over jetties, where they roll like mere pebbles.

"It is to the continual action of waves that cliffs are due, that is to say the vertical escarpments serving in some places as shore for the sea. Such escarpments are seen on the coasts of the English Channel, both in France and in England. Unceasingly the ocean undermines them, causes pieces to fall down which it triturates into pebbles, and makes its way so much farther inland. History has preserved the memory of towers, dwellings, even villages, that have had to be abandoned little by little on account of similar landslides, and that to-day have entirely disappeared beneath the waves."

"Stirred up like that, the waters of the sea are not likely to become putrid," remarked Jules.

"The movement of the waves alone would not suffice to insure the incorruptibility of sea-water. Another cause of salubrity comes in here. The waters of the sea hold in solution numerous substances that give it an extremely disagreeable taste, but prevent its corruption."

"Then you cannot drink sea-water?" Emile asked.

"No, not even if you were pressed with the greatest thirst."

"And what taste has sea-water?"

"A taste at once bitter and salt, offensive to the palate and causing nausea. That taste comes from the dissolved substances. The most abundant is ordinary salt, the salt we use for seasoning our food."

"Salt, however," objected Jules, "has no disagreeable taste, although one cannot drink a glass of salt water."

"Doubtless; but in the waters of the sea it is accompanied by many other dissolved substances whose taste is very disagreeable. The degree of salt varies in different seas. A liter of water in the Mediterranean contains 44 grams of saline substances; a liter of water in the Atlantic Ocean contains only 32.

"An attempt has been made to estimate, approximately, the total quantity of salt contained in the ocean. Were the ocean dried up and all its saline ingredients left at the bottom, they would suffice to cover the whole surface of the earth with a uniform layer ten meters thick."

"Oh, what a lot of salt!" cried Emile. "We should never see the end of it, however much we salted our food. Then salt is obtained from the sea?"

"Certainly. A low, level stretch of seashore is selected, basins are dug, shallow but of considerable extent; these are called salt marshes. Then the sea-water is admitted to these basins. When they are full, the communication with the sea is closed. The work on salt marshes is done in the summer. The heat of the sun causes the water to evaporate little by little, and the salt remains in a crystalline crust that is removed with rakes. The accumulated salt is piled up in a big heap to let it drain."

"If we should put a plate of salt water in the sun, would that be doing in a small way what is done in the salt marshes?" asked Jules.

"Exactly: the water would disappear, evaporated by the sun, and the salt would remain in the plate."

"There are lots of fish in the sea, I know," said Claire, "small, large, and monstrous. The sardine, cod, anchovy, tunny-fish, and ever so many more come to us from the sea. There are also mollusks, as you call them, also animals that cover themselves with a shell; then enormous crabs with claws bigger than a man's fist; and a lot of other creatures that I don't know. What do they all live on?"

"First, they eat one another a good deal. The weakest becomes the prey of a stronger one, which in its turn finds its master and becomes food for it. But it is plain that if the inhabitants of the sea had no other resource than devouring one another, sooner or later nourishment would fail them and they would perish.

Seaweed

"Therefore, in this matter of nutrition, things are ordered in the sea much as they are on land. Plants furnish alimentary matter. Certain species feed on the plant, others devour those that eat the plant; so that, directly or indirectly, vegetation really nourishes them all."

"I understand," said Jules. "A sheep browses the grass, a wolf eats the sheep, and so it is the grass that nourishes the wolf. There are, then, plants in the sea?"

"In great abundance. Our prairies are not more grassy than the bottom of the sea. Only, marine plants differ much from land ones. They never have blossoms, never anything that can be likened to leaves, never any roots. They attach themselves to rocks by a stickiness at their base, without being able to draw nourishment from them. They feed on water and not on the soil. Some resemble sticky thongs, folded ribbons, long manes; others take the form of little tufted buds, soft top-knots, wavy plumes; still others are slashed in strips, rolled in spirals, or shaped like coarse, slimy threads. Some are olive-green, or pale rose-color; others are honey-yellow, or bright red. These odd plants are called seaweeds."

CHAPTER LXXV

RUNNING WATERS

"I HAVE BEEN told," said Emile, "that the Rhone empties its waters into the sea."

"The Rhone does run into the sea," returned his uncle. "It pours into it every second five million liters of water."

"Receiving so much water continually, does not the sea end by overflowing, like a basin, when it is too full?"

"You are out in your reckoning, my dear child. The Rhone is not the only river that goes to the sea. In France alone there are the Garonne, Loire, Seine, and many less important ones. And that is only a very small part of the streams that flow into the sea. All the rivers in the world join it, absolutely all. The Amazon, in South America, is 1400 leagues long, and ten leagues wide at its mouth. What an immense quantity of water it must furnish!

"Imagine that all the streams in the world, small as well as large, the tiniest brooks no less than the enormous rivers, flow unceasingly into the sea. You know the little brook with the crabs. In certain places Emile can jump across it; scarcely anywhere is the water over his knees. Well, the brook goes to the sea exactly as the Amazon does; every second it casts its few liters of water into it; that is all it can do. But it does not dare, tiny little stream, to make the voyage alone and go and find the sea, the immense sea, all by itself. It meets company on the way, joins its thread of clear water to stronger streams which become rivers by joining their forces; the sea-going-river receives tributary streams, and the sea, in receiving the river, drinks the tiny brook."

"All running waters," said Jules, "brooks, torrents, streams, rivers, run into the sea without a break, and that takes place all over the world, so that every second the sea receives incalculable volumes of water. So I come back to Emile's question: How is it that, continually receiving so much water, the sea does not overflow?"

"If, when full, a reservoir receives from a spring just as much as it lets out through some opening, can this reservoir overflow, even when water is always coming in?"

"Certainly not: losing as much as it receives, it must always keep the same level."

"It is the same with the sea. It loses just as much as it gains, and therefore its level always remains the same. Brooks, torrents, streams, rivers, all run into the sea; but brooks, torrents, streams, and rivers also come from the sea. They carry back to the immense reservoir what they took from it, and not a drop more."

"If the crab brook comes from the sea," interposed Emile, "as you say, its water ought to be salt; but I know very well it is not, in the least."

"Certainly it is not salt; but the brook does not come out of the sea as the water of a ditch comes from a reservoir. In coming from the sea, before becoming what it is, the brook has first passed through the air as clouds."

"As clouds?"

"As clouds, my little friend. Let us recall something I told you a while ago.

"The heat of the sun causes water to evaporate; it reduces it to something invisible, to vapor that is dissipated in the air. Seas present a surface three times that of the dry land. Over these immensities there is constantly taking place an enormous evaporation, raising into the air a part of the waters of the sea. The vapor thus formed becomes clouds; the clouds are borne in all directions, letting down snow and rain; this rain and melted snow penetrate the ground, filter down and give birth to springs, which gradually, by their union, become brooks, streams, and rivers."

"I see why the water of brooks is not salt," said Jules, "although it comes from the sea. When you put salt water in a plate in the sun, only the water goes away; the salt remains. The vapor that rises from the sea is not salt, because the salt does not go with it when it forms. So streams fed by snow and rain that fall from the clouds cannot be salt."

"What you have just told us is very remarkable, Uncle," observed Claire. "All water-courses, rivers, streams, torrents, brooks, come from and return to the sea."

"They come from the sea, an inexhaustible reservoir that covers with its waters a surface three times larger than that of all the continents joined together; from the sea, whose abysses go down at some places to the depth of 14 kilometers, and receive unceasingly the tribute of all the water-courses of the world, without ever being taxed beyond their capacity. The enormous surface of the sea furnishes the air with vapor which turns into clouds; later these clouds dissolve in rain and, chased by the wind, travel like immense watering-pots over the ground, rendering it fertile. In their turn, rain and snow, precipitated by the clouds, give birth to the rivers that carry their waters to the sea. In that way a continual current is effected which, starting from the sea, returns to the sea, after having traveled through the atmosphere in the form of clouds, watered the earth as rain, and crossed continents as rivers.

"The sea is the common reservoir of the waters. Rivers, springs, fountains, every little brooklet, all come from and all return to it. The water of a dewdrop, the water that circulates in the sap of plants, the water that forms beads of perspiration on our foreheads, all come from the sea and are on their way back to it. However small the little drop, do not fear that it will lose its way. If the arid sand drinks it up, the sun will know how to draw it out again and send it to rejoin the vapor in the atmosphere and, sooner or later, to reënter the ocean-basin. Nothing is lost, nothing escapes the eye of God, who has measured the oceans in the hollow of His hand, and knows the number of their drops of water."

CHAPTER LXXVI

THE SWARM

UNCLE PAUL was still talking when they heard a persistent noise in the garden: pom! pom! pom! pom! as if some smith had set up his anvil under the big elder-tree. They ran to see what it was. Jacques was gravely tapping with a key on the watering can: pom! pom! pom! pom! Mother Ambroisine was busily beating a copper saucepan with a small stone: pom! pom! pom! pom!

Have our two good servants lost their heads, that they are giving themselves up, with the most serious air in the world, to this charivari? Without suspending their singular occupation, they exchange a few words. "They are going toward the currant-bush," says Jacques. "They look as if they were going away," answers Mother Ambroisine; and the pom! pom! pom! pom! is resumed.

Just then Uncle Paul and his nephews and niece come up. One glance is enough to explain everything to Uncle Paul. Over the garden there is a kind of red smoke flying, which sometimes rises and sometimes sinks, sometimes scatters and sometimes comes together in a compact mass. A monotonous whirring of wings proceeds from the midst of the red smoke. Jacques and Mother Ambroisine, still tapping, follow the cloud. Uncle Paul looks on, greatly preoccupied. Emile, Jules, and Claire look at each other, surprised at what is going on.

The little cloud descends, it approaches the currant-bush, as Jacques had foreseen, passes around it, examines it, chooses a branch. And now pom! pom! pom! pom! louder than ever. On the branch selected a round mass is formed, visibly increasing while the cloud, less and less compact, whirls around. Jacques and Mother Ambroisine stop tapping. Soon there hangs from the branch of the currant-bush a large bunch, from which the last comers of the living cloud depart to return an instant later. All is over; one can now approach.

Emile, who suspects it is bees, would like to return to the house. His old misadventure with the hive has left him with lively remembrances. To reassure him his uncle takes him by the hand. Emile bravely approaches the currant-bush. What risk can he run with his uncle? Jules and Claire come close also; it is worth the trouble.

Now, on the currant-bush hangs a bunch of bees, all close together. Some belated ones come from here and there, choose a good place, and cling on to the preceding ones. The branch bends under the burden, for there are several thousands on it. The first arrivals, doubtless the most robust, since they will have to support the whole load, have seized the branch with the claws of their forefeet; others have come and fastened themselves to the hind feet of the first bees, and in their turn have served as suspension points to a third rank; then, gradually, to a fourth, fifth, sixth, and more still, meantime diminishing in number, until finally they are all clinging there by their hands, as one might say. The children stand in wonder before the bunch of bees, whose red down and lustrous wings shine in the sun; but they prudently keep at a distance.

"Do we not run the risk of being stung by getting so near?" Jules asked.

"In their present condition bees rarely make use of their sting. If you foolishly went and tormented them, I would not answer for their conduct; but leave them alone, and you can watch them at your ease, without any fear. They have other cares now than thinking of stinging little curious boys!"

"And what cares? They look very peaceful; one would say they were all asleep."

"The grave cares of a people who have no country and seek to create one for themselves."

"Bees have a country, then?"

"They have a hive, which amounts to the same thing for them."

"Then they are looking for a hive to live in?"

"They are looking for a hive."

"And where do these homeless bees come from?"

"They come from the old hive in the garden."

"They might have stayed there, instead of going out to seek their fortunes."

"They could not. The population of the hive increased, and there was not room enough for all. So the most adventurous, under the guidance of a queen, expatriated themselves to found a colony elsewhere. The emigrating troop is called a swarm."

"The queen who leads the swarm—she must be there in the common bunch?"

"She is. It is she who, alighting on the currant-bush, determined the halt of the entire company."

These words, country, queen, emigrants, colony, had impressed the children's imaginations; they were astonished to hear the terms of human politics applied to bees. Questions came one after another, but Uncle Paul turned a deaf ear.

"Wait until the swarm is gathered into the hive, and I will tell you at length the splendid story of the bees. At present I will only answer Claire's question as to why Jacques and Mother Ambroisine tapped on the watering-pot and the saucepan.

"If the swarm had flown off into the country, it would have been lost to us. It was necessary to induce it to alight on a tree in the garden and there form itself into a bunch. It has always been thought that this result could be obtained by making a noise. Thus the sound of thunder is imitated and, as it is said, the bees, afraid of the perils of an approaching storm, quickly seek refuge. I do not believe bees are silly enough to fear a storm because of this tapping on an old pot. They alight where they please, when they please, and not far from the old hive, provided the place suits them."

Jacques, with a saw in one hand and a hammer in the other, called to Uncle Paul. With some new boards he was going to make a house for the swarm. By evening the hive was ready. At the bottom were three little holes for the bees to go in and out, and inside some pegs for holding the future honey-combs. A large flag-stone had been placed against the wall for the hive to stand on. At night-fall they went to the currant-bush. The bunch of bees was put into the hive, and a few shakes detached it from the branch. Finally the hive was put in place on its support.

The next morning Jules watched to see what the bees were doing. The house had suited them. They were to be seen coming, one by one, out of the little doors of the hive, rubbing themselves a moment in the sun on the flag-stone, and then flying away to the flowers in the garden. They were at work. The colony was founded. At a grand council they had decided matters during the night.

CHAPTER LXXVII

WAX

IT WAS not necessary to remind Uncle Paul of his promise. He took advantage of the first leisure moment to tell the children the story of the bees.

"A well-peopled hive contains from twenty to thirty thousand bees. That is about the population of our secondary towns. In a town all cannot follow the same trade. Bakers make bread, masons houses, carpenters furniture, tailors clothes; in short, there are artisans for every occupation. In like manner, in the social economy of the beehive, there are various divisions; namely, that of the mothers, that of the fathers, and that of the workers.

"For the first, there is only one bee in each hive. This bee, mother of the whole population, is called the queen. She is distinguished from the workers by a large body and the absence of working implements. Her business is to lay eggs. She has as many as twelve hundred at a time in her body, and others keep on forming as fast as the first are laid. What a formidable business is the queen's! But then, what respectful attentions, what tender care the other bees show to their common mother! They feed the noble mother by the mouthful; they give her of the best, for she has not time to gather for herself, and, to tell the truth, would not know how to do it if she had. To lay and lay is her one and only function.

DRONE

"The business of father falls to six or eight hundred idlers called drones. They are larger than the workers and smaller than the queen. Their large bulging eyes join together on the top of the head. They have no sting. Only the queen and the workers have the right to carry the poisoned stiletto. The drones are deprived of this weapon. One asks, what use are they? One day they form a retinue of honor to the queen, who takes a fancy to fly through the air; then hardly anything more is heard of them. They perish miserably in the open, or, if they return to the hive, are coldly received by the workers, who look at them unkindly for exhausting the provisions without ever adding to them. At first they treat them to some smart blows to show them that idlers are not wanted in a working society; and if they fail to understand, a resolution is taken. One fine morning they kill every one of them. The bodies are swept out of the hive, and that's the end of it.

"Now come the workers, about twenty or thirty thousand bees to one queen. These are called working-bees. They are the ones you see in the garden flying from one flower to another, gathering the harvest. Other workers, a little older and consequently more experienced, remain in the hive to look after the housekeeping and to distribute nourishment to the nurslings

WORKER

hatched from the eggs laid by the queen. There are, then, two bodies of workers to be distinguished: the wax-bees, younger, which make wax and gather the materials for honey; the nurses, older, which stay at home to bring up the family. These two kinds of workers are not mutually exclusive. When young, full of ardor, adventurous, the bee follows the trade of wax-maker. It goes to the fields, seeking viands, visits the flowers, or sometimes is forced to assert itself and unsheath its sting, to put to flight some evil-intentioned aggressor; it sweats wax to make the storehouse and the little rooms where the brood of young ones is kept. Growing older, it gains experience, but loses its first ardor. Then it stays at home, turns nurse, and occupies itself with the delicate task of rearing the young."

This preamble of Uncle Paul's, defining the three industrial classes of the bees, appeared to interest the children greatly, and they were surprised to find that insects have such marvelously elaborate social laws. At the very first opportunity Jules began questioning his uncle. The impatient child wanted to know everything at once.

"You say the wax-bees make wax. I thought they found it ready-made in flowers."

"They do not find it ready-made. They make it, sweat it, that is the word, as the oyster sweats the stone of his shell, as the meleagrina sweats the substance of its mother-of-pearl and its pearls.

"If you look closely at a bee's stomach, you will see it is composed of several pieces or rings fitting into each other. The stomach of all insects has, moreover, the same formation. This arrangement of several parts fitted endwise is found in the horns or antennæ, as well as in the legs, of all insects without exception. It is precisely to this division into separate pieces fitted endwise that the word insect alludes, its meaning being cut in pieces. Is not the body of an insect composed, in fact, of a series of pieces placed end to end?

"Let us come back to the bee's stomach. In the fold separating one ring from the next there is found, underneath, in the middle of the stomach, the wax-producing mechanism. There, little by little, the waxy matter oozes out, just as with us sweat oozes through the skin. This matter accumulates in a thin layer which the insect detaches by rubbing the stomach with its legs. There are eight of these wax-producers. When one is idle, another is working; so that the bee always has some layer of wax at its disposal."

"And what does the bee do with its wax?"

"It builds cells, that is to say storehouses, where the honey is preserved, and little rooms where the young bees in the form of larvæ are raised."

"It builds its house, then," put in Emile, "with the layers of wax taken from the folds of its stomach. And there, you see, the bee shows a very original and inventive mind. It is as if, in order to build a house, we should rub our sides so as to get from them the blocks of cut stone we needed."

"The snail," concluded Uncle Paul, "has already accustomed us to these original ideas of animals. It sweats the stone for its shell."

CHAPTER LXXVIII

THE CELLS

" IN ORDER to store the supply of honey and lodge the larvæ, the bees build with their wax little rooms called cells, open at one end and closed at the other. They are six-sided and arranged with perfect regularity. In geometrical terms, each would be called a hexagonal prism, or a prism with six facets.

"Do not be surprised at this introduction of terms belonging to the beautiful and severe science of form—of geometry, in short. Bees are geometricians of consummate skill. Their constructions have required the exercise of the highest intelligence. All the power of human reason was necessary to follow, step by step, the insect's science. I will return presently to this fine subject, a very difficult one, but I will try to make it intelligible to you.

"The cells are placed horizontally, back to back and end to end, in pairs, with the closed ends joining. Furthermore, they are arranged side by side in greater or less number, and they touch each other by their flat faces, each one of which serves as partition wall for two contiguous cells. The two layers of cells, back to back at their closed ends, constitute what is called a comb or honey-comb. On one side of this comb are found all the entrances to the cells of the corresponding layer; on the other side the cells of the second layer open. Finally, the honey-comb is suspended vertically in the hive, with half its openings to the right and half to the left. It adheres by its upper edge to the roof of the hive, or to the bars that cross it inside.

"One comb is not enough when the population is numerous; others are constructed like the first. The various combs, ranged parallel to one another, leave free intervening spaces. These are the streets, the public squares, the thoroughfares, on which the

273

openings of the two layers of cells belonging to neighboring combs give, as the doors of our houses open on the right and left of a street. There the bees circulate, going from one door to another to deposit their honey in the cells used as storehouses, or to distribute nourishment to the young larvæ lodged one by one in other cells. In these same public places they assemble when necessary, hold consultations, and deliberate on the affairs of the community. There, for example, among the nurses going from door to door to see whether the larvæ need feeding, and the wax-bees rubbing themselves vehemently to extract the wax and begin to build, is plotted the extermination of the drones; there, when the birth of a new queen menaces the hive with civil war, the project of emigration ripens. There—But let us not anticipate. Let us return to the cells."

"I am longing to know the whole of the strange story of the bees," Jules broke in.

"Patience! First of all let us see how the cells are constructed. The bee that feels that it is supplied with the materials for making wax rubs itself and extracts a sheet of wax from the folds of its rings. With the little layer of wax between its teeth, that is to say between its two mandibles, it squeezes through the press of its comrades. 'Let me pass,' it seems to say; 'see, I have something to work with.' The crowd makes way. The bee takes its place in the middle of the workyard. The wax is kneaded between its mandibles, pounded to pieces, then flattened out into a ribbon, pounded again, and once more kneaded into a compact mass. At the same time it is impregnated with a kind of saliva that gives it flexibility. When the material is at the proper stage, the bee applies it bit by bit. To cut off the surplus, the mandibles serve as scissors; the antennæ, in continual motion, serve it as probe and measuring-compasses; they feel the wall of wax to judge of its thickness; they plunge into the cavity to find out its depth. What exquisite touch in this pair of living compasses, to bring to successful completion a construction so delicate and regular! Moreover, if the worker is a novice, master-bees are there to watch it with an experienced eye, to seize on the slightest fault at once and hasten to remedy it. The maladroit worker modestly steps aside and watches in order to learn. The trick learned, it sets to work again. With thousands of wax-bees working together, a comb two or three decimeters wide is often a day's work."

"You told us," said Claire, "that the cells are especially remarkable for their geometrical arrangement."

"I am just coming to that magnificent topic, but I shall treat it briefly, I warn you. You are far from being able to follow yet in its superior beauties the architecture of the bees. Yes, my dear Jules, the wax house of a poor insect, to be well understood, demands knowledge that very few persons possess. Ah, you may study ever so long before you are able fully to understand this marvel! For the present, here is what I will tell you.

"The cells serve, some as store-rooms for the honey, others as nests for the little ones. They are made of wax, a material that the bees cannot procure in indefinite quantities. They must wait until the stomach sweats a little layer of it, and it forms very slowly, at the expense of the insect's very substance. The bee builds with the materials of its own body, it impoverishes itself in sweating the wherewithal to construct the cells. You can judge from that how precious a thing wax is to the bees, and with what strict economy they must use it.

"And yet the innumerable family must be lodged, honey store-rooms must multiply to supply the wants of the community. Moreover, it is necessary that these store-rooms and nurseries take up as little room as possible, so as not to encumber the hive, and to permit free circulation to the twenty or thirty thousand inhabitants of the city. In fine, one of the hardest problems is presented to the bees: they must make the greatest possible number of cells in the least space and with the least wax possible. Well, friend Jules, do you think you could solve the bees' problem?"

"Alas! Uncle, I hardly understand the statement of it."

"To economize the wax, a very simple way suggests itself at the outset: it is to make the partitions of the cells very thin. You may be quite sure the bees are equal to this elementary requirement. They make the wax walls scarcely as thick as a sheet of paper. But that is not enough: it is necessary above all to take the form into consideration and to seek the most economical shape. Let us try. What shape shall we give the cells to satisfy the conditions of economy in space and wax?

"First of all let us suppose them to be round. Let us trace on paper some circles of equal size and touching one another. Between three of these contiguous circles there will always be an unoccupied space. The round form will not do, then, for the cells, since there will always be a waste of space, or empty intervals.

"Let us make them square. We will trace equal squares on the paper. In going about it properly we can arrange the squares side by side without leaving any empty spaces between them. Look at the inlaid floor of this room, composed of little square red bricks. These bricks leave no intervening spaces; they touch on every side. The square form, therefore, suits the first condition, namely: to utilize all the space.

"But here is where another difficulty arises. Cells fashioned on the square model would not hold enough honey for the quantity of wax used in constructing them. In order to increase their capacity, you must increase as much as possible the number of their facets. I will not try to demonstrate to you this beautiful truth; it is beyond your intelligence. Geometry affirms it; let us consider it a fact.

"Starting from that, the choice is soon made. Among all the regular figures that can be placed side by side without leaving an unoccupied space, you must choose that which has the greatest number of sides, for that is the one that will hold the most honey for the same quantity of wax used.

"Geometry teaches that the only regular figures that can be arranged without waste of space are: the three-sided figure, or triangle; the four-sided, or square; and the six-sided, or hexagon. That is all: no other regular figures touch all around so as to leave no empty spaces between them.

"So it is, then, in the hexagonal form, or form with six sides, that the cells can occupy, collectively, the least space, use the least wax, and hold the most honey. Bees, knowing these things better than any one else, make hexagonal cells, never any other kind."

"Then bees have reason," remarked Claire, "like ours; even

superior, if they can solve such problems?"

"If bees constructed their cells after a premeditated, considered, calculated plan, it would be something alarming, my dear child: animals would rival man. Bees are profound geometricians because they work, unconsciously, under the inspiration of the sublime Geometrician. Let us stop this talk, which I fear you have not wholly understood; but, at any rate, I have opened your eyes to one of the greatest wonders of this world."

CHAPTER LXXIX

HONEY

"THE BEE is diligent: at sunrise it is at work, far from the hive, visiting the flowers one by one. You already know what it is in flowers that attracts insects: I have told you about the nectar, that sweet liquor that oozes out at the bottom of the corolla to entice the little winged people and make them shake the anthers on the stigma. This nectar is what the bee wants. It is its great feast, the great feast also of the little ones and the queen-mother; it is the prime ingredient of honey. How carry home a liquid so that others may enjoy it? The bee possesses neither pitcher, jar, pot, nor anything of the sort. I am wrong: like the ant that carries the plant-lice's milk to the workers, it is provided with a natural can, stomach, paunch, or crop.

"The bee enters a flower, plunges to the bottom of the corolla a long and flexible trunk, a kind of tongue that laps the sweet liquor. Droplet by droplet, drawn from this flower and that, the crop is filled. The bee at the same time nibbles a few grains of pollen. Moreover, it proposes to carry a good load of it to the hive. It has special utensils for this work: first, the down of its body, then the brushes and baskets that its legs supply. The down and the brushes are used for harvesting; the baskets for carrying.

"First the bee rolls delightedly among the stamens to cover itself with pollen. Then it passes and re-passes over its velvety body the extremities of its hind legs, where is found a square piece bristling on the inside with short and rough hairs which serve as a brush. The grains of pollen scattered over the down of the insect are thus gathered together into a little pellet, which the intermediary legs seize in order to place it in one or other of the baskets. They call by this name a hollow edged with hair on the outside of the hind legs, a little above the brushes. It is there the pellets of pollen are piled up as fast as the brushes gather them on the powdery down. The load does not fall, because it is held by the hairs that edge the basket; it is also stuck against the bottom. The queen and the drones have not these working implements. Utensils are useless to those who do not work."

"The little yellow masses one sees on the hind legs of bees visiting the flowers are loads of pollen contained in the baskets?" asked Jules.

"Exactly. The bee has lapped so much sweet from the corollas, has brushed its pollen-powdered sides so often, that finally the crop is full and the baskets are running over. It is time to go back to the hive, time for a flight made heavy with so much treasure.

"Let us take advantage of the time used in the return journey to inform ourselves about the origin of honey. The bee carries with it a sugary liquor in its crop, two balls of pollen in its baskets; but all that is not yet honey. Real honey the bee prepares with the ingredients that we have just seen it gather; it cooks it, lets it simmer in its crop. Its little stomach is better than a real pot for carrying: it is an admirable alembic, in which the liquid that has been lapped up and the grains of pollen that have been nibbled are worked by digestion and converted into a delicious marmalade, which is honey. This skilful cooking finished, the content of the crop is honey.

"The bee arrives at the hive. If by good fortune the queen-mother is encountered, the workman does reverence to her and offers her, from mouth to mouth, a sip of honey, the first from its crop. Then it seeks an empty cell, inserts its head into the storeroom, projects its tongue, and spits out the contents of its stomach; and there you have real honey disgorged by the bee."

"Is it all disgorged?" Emile asked.

"Not all. The crop's contents are usually divided into three parts: one for the nurses that remain in the hive to do the house-work; a second for the little ones still in the nest; a third kept by the bee that has prepared the honey. Must it not have food in order to work well?"

"Then bees feed on honey?"

"Without a doubt. You imagined perhaps that bees made

honey expressly for man. Undeceive yourself: bees make honey for themselves and not for us. We plunder their riches."

"What becomes of the little balls of pollen?" inquired Jules.

"The pollen enters into the making of honey, and serves as nourishment for the bees. The working bee, on its return from harvesting, puts its hind legs into a cell where there is neither larva nor honey, and with the end of its middle legs it detaches the pellets and pushes them to the bottom. In repeating its trips it ends by filling both the cell in which the honey is disgorged and that in which the pollen is stored. The nurses draw on these provisions when they go from cell to cell, distributing small portions to the little ones; thence also they get their own food; in fact, the whole population finds its resources there when bad weather comes.

"Flowers do not last all the year, and, moreover, there are days of rest, rainy days when the bees cannot go out. It is necessary, therefore, to have pollen and honey in reserve, and to have a good supply. So, when flowers are plenty and the harvest exceeds immediate requirements, the workers gather honey and pollen untiringly and store it in cells, which they close, as soon as full, with a cover of wax.

"These are reserve supplies, safeguards for the future in case of scarcity. The wax cover is religiously respected; it would be a state crime to touch it prematurely. In time of want the seals are removed and each one draws from the open comb, but with restraint and sobriety. The comb exhausted, they break the seals of another."

"How are young bees fed?" was Jules's next question.

"When the cells destined to serve as nests are prepared in sufficient number by the wax-bees, the queen-mother goes from one to another, dragging with much effort her fruitful womb. The nurses form a respectful retinue. One egg, one only, is laid in each cell. In a few days—from three to six—there comes from this egg a larva, a little white worm, without legs, bent like a comma. Now begins the nurses' delicate work.

"They must every day, and several times a day, distribute nourishment to the little worms, not honey or pollen in its natural state, but a preparation of increasing strength such as delicate stomachs need at first. It is, in the beginning, a liquid paste, almost tasteless; then something sweeter; and finally pure honey, nourishment at its full strength. Do we offer a slice of beef to a crying baby? No, but milk first and then pap. Bees do the same: they have honey, strong food, for the strong; and weaker nourishment, tasteless pap, for the weak. How do they prepare these more or less substantial foods? It would be hard to say. Perhaps they mix pollen and honey in different proportions.

"In six days the larvæ, called brood-comb, have attained their development. Then, like the larvæ of other insects, they retire from the world to undergo metamorphosis. In order to protect its suffering flesh at the critical moment of its transfiguration, each larva lines the inside of its cell with silk, and the working-bees close the cell with a cover of wax. In the silk-lined case the skin is cast off and the passage to the state of nymph accomplished. Twelve days later the nymph awakes from the deep sleep of the second birth; it shakes itself, tears its narrow swaddling-clothes, and comes forth a bee. The wax cover is gnawed by the inclosed insect as well as by the working-bees lending a ready hand to the resuscitated; and the hive counts one more citizen. The new-born bee makes its toilet a little, dries its wings, polishes its body, and is off to work. It knows its trade without having had to learn it: wax-bee in its youth, nurse in its old age."

CHAPTER LXXX

THE QUEEN BEE

"THE EGGS destined to give birth to queens are laid in special cells, much more spacious and solid than those where the working-bees hatch. Their shape is, in a general way, that of a thimble. They are fastened to the edge of the combs and are called royal cells."

"When she lays in a large or small cell," asked Jules, "does the queen know whether the egg is that of a queen or of a working-bee?"

QUEEN BEE

"She does not know, she does not need to know. There is no difference between the queen-eggs and working-bee-eggs. Its treatment alone decides the issue for the egg. Treated in a certain manner, the young larva becomes a queen, on whom depends the future prosperity of the hive; treated in another way, it becomes one of the working people and is furnished with brushes and baskets. Bees make their queens at will; the first egg laid would suffice to fill the royal functions worthily, if treated with that end in view. And what does not treatment, or education, accomplish with us in our tender years? It does not make us kings or peasants, but honest people, which is better; and scoundrels, which is worse.

"It need not be said that the bees' pedagogic methods are not the same as ours. Man, as much mind as matter, if not more, turns his attention above all to the generous impulses of the heart, the noble aspirations of the soul. With bees education is purely animal, and is governed by the dictates of the belly. The kind of food makes either the queen or the working-bee. For the larvæ that are to discharge the functions of royalty the nurses prepare a special pap, a royal dish of which only they know the secret. Whoever eats of it is consecrated queen.

"This strengthening nourishment brings about a greater development than usual; for that reason, as I told you, the larvæ destined for royalty are lodged in spacious cells. For these noble cradles wax is used with prodigality. No more hexagonal, parsimonious forms, no thin partitions; a large and sumptuously thick thimble. Economy is silent where queens are concerned."

"It is, then, without the actual queen's knowledge that bees make other queens?"

"Yes, my friend. The queen is excessively jealous, she cannot endure in the hive any bee whose presence may bring the slightest diminution to her royal prerogatives. Woe to the pretenders that should get in her way! 'Ah! you come to supplant me, to steal from me the love of my subjects!' Ah, this! Ah, that! It would be something horrible, my children. Read the history of mankind, and you will see what disasters crowned heads, brought to bay, can inflict upon nations. But the working-bees are strong-minded, they know that nothing lasts in this world, not even queens. They treat the reigning sovereign with the greatest respect, without losing sight of the future, which demands other queens. They must have them to perpetuate the race; they will have them, whether or no. To this end the royal pap is served to the larvæ in the large cells.

"Now, in the spring, when the working-bees and drones are already hatched, a loud rustling is heard in the royal cells. They are the young queens trying to get out of their wax prisons. The nurses and wax-bees are there, standing guard in a dense battalion. They keep the young queens in their cells by force; to prevent their getting out, they reinforce the wax inclosures, they mend the broken covers. 'It is not time for you to show yourselves,' they seem to say; 'there is danger!' And very respectfully they resort to violence. Impatient, the young queens renew their rustling.

"The queen-mother has heard them. She hastens up in a passion. She stamps with rage on the royal cells, she sends pieces of the wax covers flying and, dragging the pretenders from their cells, she pitilessly tears them to pieces. Several succumb under her blows; but the people surround her, encircle her closely, and little by little draw her away from the scene of carnage. The future is saved: there are still some queens left.

"In the meantime wrath is excited and civil war breaks out. Some lean to the old queen, others to the young ones. In this conflict of opinions disorder and tumult succeed to peaceful activity. The hive is filled with menacing buzzings, the well-filled storehouses are given over to pillage. There is an orgy of feasting with no thought of the morrow. Dagger-thrusts are exchanged. The queen decides on a master-stroke: she abandons the ungrateful country, the country that she founded and that now raises up rivals against her. 'Let them that love me follow me!' And behold her proudly rushing out of the hive, never to enter it again. Her partizans fly away with her. The emigrating troop forms a swarm, which goes forth to found a new colony elsewhere.

"To restore order, the working-bees that were away during the tumult come and join the bees left in the hive. Two young queens set up their rights. Which of them shall reign? A duel to the death shall decide it. They come out of their cells. Hardly have they caught sight of each other when they join in shock of battle, rear upright, seize with their mandibles each an antenna of the other, and hold themselves head to head, breast to breast. In this position, each would only have to bend the end of its stomach a little to plunge its poisoned sting into its rival's body. But that would be a double death, and their instinct forbids them a mode of assault in which both would perish. They separate and retire. But the people gathered around them prevent their getting away: one of them must succumb. The two queens return to the attack. The more skilful one, at a moment when the other is off guard, jumps on its rival's back, seizes it where the wing joins the body, and stings it in the side. The victim stretches its legs and dies. All is over. Royal unity is restored, and the hive proceeds to resume its accustomed order and work."

"The bees are very naughty to force the queens to kill one

another until there is only one left," commented Emile.

"It is necessary, my little friend; their instinct demands it. Otherwise civil war would rage unceasingly in the hive. But this hard necessity does not make them forget for one moment the respect due to royal dignity. What is to prevent their getting rid of the superfluous queens themselves, even as they so unceremoniously get rid of the drones? But this they are very careful not to do. What one of their number would dare to draw the sword against their sovereigns, even when they are a serious encumbrance? The saving of life not being in their power, they save honor by letting the pretenders fight it out among themselves.

"There is always the possibility that the queen, at a time when she is reigning alone and supreme, may perish by accident or die of old age. The bees press respectfully around the deceased; they brush her tenderly, offer her honey as if to revive her; turn her over, feel her lovingly, and treat her with all the regard they gave to her when alive. It takes several days for them to understand, at last, that she is dead, quite dead, and that all their attentions are useless. Then there is general mourning. Every evening for two or three days a lugubrious humming, a sort of funeral dirge, is heard in the hive.

"The mourning over, they think about replacing the queen. A young larva is chosen from those in the common cells. It was born to be a wax-bee, but circumstances are going to confer royalty upon it. The working-bees begin by destroying the cells adjacent to the one occupied by the sacred larva, the queen that is to be by unanimous consent. The rearing of royalty requires more space. This being secured, the remaining cell is enlarged and shaped like a thimble, as willed by the high destiny of the nursling it contains. For several days the larva is fed with royal paste, that sugary pap that makes queens, and the miracle is accomplished. The queen is dead, long live the queen!"

"The story of the bees is the best you have told us," declared Jules.

"I think so too," his uncle assented; "that is why I kept it till the last."

"What—the last?" cried Jules.

"You are not going to tell us any more stories?" asked Claire.

"Never, never?" Emile put in.

"As many as you wish, my dear children, but later. The grain is ripe, and the harvest will take up my time. Let us embrace, and finish for the present."

Since Uncle Paul, occupied with his duties in the harvest-field, no longer tells stories in the evening, Emile has gone back to his Noah's Ark. He found the hind and the elephant moldy! From the time of the story of the ants the child had suspended his visits.